BUILDING FOR ENERGY CONSERVATION

Peter Burberry

W9-BAU-051

The Architectural Press Ltd, London
Halsted Press Division
John Wiley & Sons, New York

CONTENTS

First published in book form in 1978 by The Architectural Press Ltd: London
Reprinted 1979
ISBN 0 85139 187 7 (British edition)

Published in the U.S.A. by
Halsted Press, a Division of
John Wiley & Sons, Inc.,
New York

British Library Cataloguing in Publication Data
Burberry, Peter.
  Building for energy conservation.

  'A Halsted Press book.'
  Includes index.
  1. Buildings—Energy conservation. I. Title.
TJ163.5.B84B87 1978   696   77–17943
ISBN 0–470–99350–2 (U.S.A. edition)

Printed litho in Great Britain by
W & J Mackay Limited, Chatham

1 THE ENERGY CRISIS

1.1 The unperceived crisis

For many years now the environmentally conscious have been warning us that natural resources are fast disappearing. Those more politically conscious have cautioned us that the OPEC countries (Organisation of Petroleum Exporting Countries) would, as demand for and dependence on oil increased, sharply raise their prices. It is astonishing, therefore, to have observed the whole political establishment apparently amazed by the recent oil crisis and the country totally unprepared.

Whether one believes that supplies are finally running out, that increased prices will make economic the exploitation of vast reserves of low-grade sources, or that necessity will be the mother of new energy inventions, there can be little doubt that the energy, which has been used so prolifically in buildings over the past 20 years, will be very much more expensive in the future.

It is a salutary thought that probably half of the energy used in the United Kingdom goes to heating buildings and two-thirds of this into domestic buildings. It is quite clear that in the future a fundamental aspect of the design of buildings will be the energy budget. Since energy is needed in the manufacture of materials, transport and construction as well as in environmental control, the whole problem of the functional design of buildings will become very much more important.

At present, *overall* design for energy conservation is prevented by the structure of the building design professions. In the nineteenth century, when the present pattern of professions and their responsibilities was laid down, there was no problem to be solved. With few exceptions, buildings had massive external walls and, by present standards, small windows.

Gravity circulation central heating with radiators was used if it could be afforded and open fires if not. Architects, within the constraints and design conventions of the day, did not have to think about thermal problems, while heating engineers estimated heat losses from the building but in design were concerned only with boilers, pipework and radiators. Today

the form and construction of buildings has changed dramatically, particularly in thermal terms, but the professional structure has not. The extension to St George's School, Wallasey, photograph 1.1, is well known as a building not requiring a conventional heating system, but after 15 years it remains unique. The complete avoidance of the use of a heating installation achieved in this school could not be widely applied in other buildings, but the school forms an object lesson—that cannot be ignored—of the thermal importance of the fabric of the building. The virtually total lack of follow-up can be attributed partly to the cheap oil policy which meant that gross waste of energy in buildings had no significant cost penalty, and partly to the current professional responsibilities by which architects design the form and fabric of buildings while heating engineers select boilers and plant, and size pipes and ducts. In thermal terms this division is nonsense. The shape, orientation, fenestration and materials of construction have as much, and perhaps more, effect on the thermal performance than the heating installation. Both professional education and professional scope must begin to take into account this aspect of design.

1.2 The nature of the crisis

The term 'energy crisis' has been widely and dramatically employed for nearly two years now. In most countries three different types of energy crisis are occurring simultaneously. The solutions to each of these crises are different and a great deal of confusion arises when remedies are advocated without any definition of the particular purpose they will serve.

1.21 The first crisis: acute

The first type of crisis, the immediate energy crisis, is financially based and arises because of a problem of payment for fuel. This is clearly an extremely short term problem. Insula-

1.1 *The unique Wallasey School extension which uses no conventional heating installation.*

ting buildings, for example, cannot have any very useful effect in such a short term situation. The only solution is to limit the flow of energy by controlling consumption immediately.

1.22 The second crisis: chronic

The second type of crisis derives from an increase in fuel cost which, while not imposing a problem of private or national bankruptcy, nevertheless places the design of buildings and the amount of resource devoted to energy consumption in a new light. There will be a progressive adjustment of resources toward improving features of the thermal design of buildings, improving insulation, better control systems and a wide range of other measures giving increased emphasis to building design rather than the use of plant and energy to control the environment.

1.23 The third crisis: terminal

The third type of crisis is that of the ultimate exhaustion of fossil fuels. The response to it is actively to research alternative sources of energy and patterns of consumption, and the social and population consequences which will follow. This research has no immediate effect on building design although many people, looking ahead, are endeavouring to design self-sufficient buildings.

1.24 Summary

The three types of crisis can be summarised in medical terminology (as suggested in the previous headings):

Table I Three types of energy crisis*

Type of crisis	Consequence	Action
Acute	National bankruptcy	Turn down or switch off
Chronic	High energy costs requiring adjustment in balance of use of both national and individual resources	Design buildings to be thermally efficient
Terminal	Exhaustion of fossil fuels	Develop alternative and recyclable sources of energy

1.3 Energy use in buildings
1.31 Potential for conservation

One of the fundamental facts in the tremendous out-pouring of statistics on energy is that, although the precise figure cannot be determined, virtually half of all the energy used in this country is used in buildings. It is apparent therefore that designers, owner-occupiers and all those concerned with buildings have a major responsibility for energy conservation.

Primary energy
The term primary energy defines the total amount of energy required to produce the fuel which is actually input to the building. It is clearly a better standard for comparison than the actual energy delivered which may be, according to the fuel chosen, substantially less than the energy consumed in producing and distributing the fuel.
The broad allocation of the use of primary energy in the UK is: industry 41 per cent; domestic 29 per cent; transport 16 per cent; others 14 per cent.
The figures for industry and transport include both the energy

* Readers should note that the most important tables are set in the same size as the text, so that their importance is not overlooked.

required for warming factories, garages, stations, airports etc, as well as industrial processes and vehicles. The category 'others' largely covers the warming of buildings such as shops, hospitals, schools and offices.

Useful energy
From diagram 1.2 it will be observed that, apart from economies made by means of design or operation of the building, major economies are made possible by the selection of types of fuel which have low distribution and installation losses.

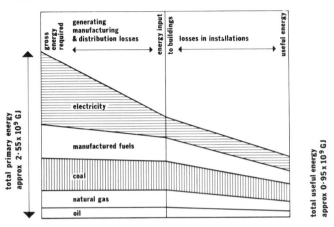

1.2 *Relationship between primary energy, energy input to buildings and useful energy for housing in the UK. (Based on data from Building Research Station Current Paper CP 56/75.)*

More than half of the total energy used in buildings goes into housing. Typical uses are: space heating 66 per cent; water heating 21 per cent; cooking 10 per cent; lighting, tv etc 3 per cent.
It is therefore in the fields of water and particularly space heating, that economies must be concentrated. There is no doubt that new buildings can be designed to use less energy and that existing buildings can be substantially improved in this respect.

1.32 Need for statutory control

As fuel prices increase building occupiers and, in due course, designers will automatically begin to take more account of economy of energy. It is important to remember, however, that while energy economy will be a major national consideration for the indefinite future, expenditure on energy by building owners, whether domestic, commercial or industrial, is only a small proportion of their individual budgets, and, in the case of speculative commercial or domestic buildings the builders responsible for the basic design will not have to pay for the heating bills. So those inevitably responsible for energy conservation decisions have insufficient self-interest to encourage good energy conserving design. Statutory control will be needed to ensure that conservation measures match the national need. Much more sophisticated measures are needed than control of U-values, and none has yet been devised.

1.4 Scope of this book

This book is concerned with the chronic phase of the energy crisis (see paragraph 1.22). It is difficult to estimate its duration; at present rates of consumption the North Sea oil reserves are thought likely to last for some 30 years and coal reserves some 300. Common sense seems to indicate that even if new reserves are discovered prices will rise as exhaustion approaches, thus limiting rates of use; and that supplies at present too expensive to exploit will become available. So it is likely that conventional fuels will play a large part in heating for the life of most existing buildings, and of those to be built in the immediate future.

2 EXAMPLES OF THERMAL DESIGN

2.1 Influence of climate

There are few places in the world where the climate allows life to take place without protection from the elements; the need for better environmental conditions is one of the major reasons for the existence of buildings. At its lowest level a building merely gives some degree of protection from rain and wind but many traditional building types in widely different climatic conditions have developed highly sophisticated thermal solutions, giving marked improvements in environmental comfort with very economical use of resources.

The courtyard houses of North Africa, the elevated wooden structures of many hot, humid climates, and the traditional English cottage have to solve very different environmental problems; they demonstrate with dramatic effectiveness how building form can control thermal performance.

2.11 Hot dry climates

A mediterranean courtyard house, photograph **2.1**, meets the problem of how to remain cool in a hot, dry climate with clear skies. The courtyards and their surrounding buildings radiate to the cold night sky and, during the night, a pool of cool air is built up in the courtyards and in the ground floor rooms. During the day the sun shines but the reservoir of heavy cool air remains for a considerable time. The walls of the building are of substantial thickness so that penetration of heat from the sun's rays on the walls takes a considerable time. Ideally the heat will reach the interior during the night, when conditions are cool and the heat entering can be dissipated by ventilation.

The walls themselves are painted white so that the minimum amount of solar heat is absorbed. In addition in hot dry climates advantage can be taken of evaporative cooling; evaporation of water takes up significant amounts of heat and gives reduced air temperatures. The fountains in the courtyard therefore have a practical value in reducing air temperatures as well as psychological attractions.

2.12 Hot humid climates

A very different problem exists in hot humid climates. There is no period of clear night sky to enable radiation to take place. The only available method for improving thermal comfort is increasing the velocity of air impinging upon occupants, both for its direct cooling effect and for the cooling which is still possible by more rapid evaporation of sweat. The type of dwelling resulting is shown in diagram **2.2**.

2.2 Raised verandah type house from a hot humid climate. The house is only one room wide with louvred walls to ensure ventilation. Overhanging roof gives outdoor sitting area.

2.1 Courtyard houses from a hot dry climate: heavyweight construction with pools of cool air in courtyard and lower rooms; and in narrow streets.

2.13 Cool climates

In higher latitudes overheating is no longer the main issue. During the winter heat input is needed and the problem becomes one of conserving the heat and giving as much habitable space as possible during the winter. Until very recently timber, peat and dung were the only fuels available and their scarcity, together with the considerable amounts of labour required for collection and delivery, meant that these fuels were expensive and could only be used sparingly. A typical traditional English cottage, with very small windows, very highly insulated roof, small volume owing to low ceilings, and a centrally sited fireplace and flue, exemplifies this.

2.2 Influence of fuel price

It is salutary to compare the designs described above with types of building which have developed during the brief periods when fuel became, for a time, more freely available.

2.21 First age of cheap fuel

The first age of cheap fuel developed, in effect, in the eighteenth century when 'sea coal' from Newcastle became freely available, if not for all, at least for the wealthy.

Many factors governed development of the Georgian houses of this age, 2.3a. The requirements of urban living limited the site, and fire regulations called for brick walls and tiled or slated roofs. Cheap energy in the form of coal made glass more freely available and the same energy source coupled with cheap labour made it possible to heat the high, well glazed rooms by separate fireplaces in each room.

The insulation value of walls and roofs was poor but it is important to remember that the terrace pattern of housing typical of the period is inherently economical of energy and that the relatively large windows were provided with shutters and heavy curtains for use at night and in bad weather.

The 'By-law' housing typical of a slightly later period has, in effect, the same thermal properties. When it was first built coal was relatively freely available even to workers, many of whom enjoyed substantial free deliveries.

2.22 Increased fuel costs

A declining economy with consequent relative increase in fuel price led to a situation where, in the late nineteenth and early twentieth centuries, only a small proportion of the area of most houses could be heated in winter. Economy curtailed aspirations for detached houses with extensive grounds and led to the development of the semi-detached house. This form, although not as economical of fuel as terraced developments,

had several features which aided its thermal performance. The living rooms were centrally placed, usually with only one external wall. At the sides ancillary accommodation such as hall and kitchens protected the living rooms, as did the bedrooms overhead, diagram 2.4. Between the houses a massive party wall with chimney breasts and chimneys provided a thermal reservoir and also contributed heat to the bedrooms.

The open fires for heating meant that very high ventilation rates were required, with a flue to carry away the smoke and adequate cracks round doors and windows to admit the flow of air required for the fire. Little thermal insulation was provided although cavity walls were used; more insulation might have given better comfort in the bedrooms. The intermittent use of an open fire would have made any saving from increased insulation highly problematic. Occupants of this type of house dressed warmly and sat in heavily upholstered furniture in range of the radiant heat from the fire.

2.23 Second age of cheap fuel

Between 1950 and 1974 a second cheap fuel era came and went. It was based upon cheap oil and, in Britain, cheap electricity. These fuels became so cheap and were so labour-saving that it was possible to equip virtually all new and a very large number of existing houses with central heating. Little consideration was given to economy of energy since at prevailing prices few people would have wished to sacrifice any amenity to save fuel, and very little capital investment in saving could be justified.

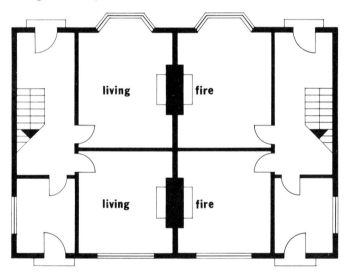

2.4 Early semis in which heating of living spaces only was planned.

*2.3a Compact Georgian house; much greater storey height and glazing areas, in lighter construction than 2.3b Workers' houses in which form, as in **a**, is based in part on freely available fuel.*

Effect on housing

Although the period is very short to have influenced house design it is possible to discern several features which can be attributed to cheap fuel. These are the large windows and the open planning which is made possible if heating is not confined to small areas; see photograph **2.6**.

Effect on larger buildings

Outside the housing field development of heating has been more dramatic. Until well into the twentieth century many non-domestic buildings were heated by open fires in the same way as houses. An increasing number were provided with simple solid fuel fired radiator systems. With the era of cheap energy and, particularly, cheap electricity it began to be possible to provide not only heating for large buildings but also mechanical ventilation and cooling. Fully glazed and high rise buildings which could not have been made habitable in previous ages could be designed and equipped with installations capable of maintaining comfort throughout the year. Dependence upon the window for ventilation and daylight, which had governed the form of buildings throughout history, was no longer the case and completely new planning concepts became possible because of the use of environmental control installations. While the use of this type of equipment clearly has great value in many circumstances, for a period many badly conceived buildings were designed which depended upon wasteful use of plant and energy to make them habitable.

Deteriorating design

It is depressing, at a time when much highly sophisticated plant is being used for environmental control, to realise that throughout much of history buildings themselves were highly developed examples of applied science, achieving high levels of performance even when designers could not define or quantify the factors that they took into account. Now that architects consciously design in thermal terms the quality of design has deteriorated and expensive plant and running costs are accepted as easy design solutions.

2.3 Significance of form, fabric and fenestration
2.31 Wallasey school

St George's School, Wallasey, Liverpool, is one of the few modern buildings where thermal performance has been considered as a factor in the design of the form and fabric of the building. One of the classroom blocks in this school has large south-facing double glazed windows, massive brick walls, and concrete roofs and floors with a very substantial amount of external thermal insulation, diagram **2.5**. This building is designed so that the fabric is able to store heat and return it to the environment when needed. An ingenious arrangement of windows in the double skin wall enables ventilation control. The school was originally used without a heating installation. The thermal system of this building is often referred to as

2.6 *A recent house, relying on general heating and plentiful fuel supplies.*

solar heating but in fact three factors contribute to the heating. First there is the solar gain through the large windows, second, heat gain from the occupants, and third, heat gain from the tungsten lighting system. The building has been occupied successfully for a number of years, and it is important to bear in mind the significance in terms of overall thermal performance of the output of heat from the occupants and from the lighting. Although children do not emit as much heat as adults in similar circumstances they tend to be rather more active, and they require a distinctly lower air temperature for comfort. In a school classroom of approximately 48 m² it is not unusual to find a population of 40 children. The area per pupil thus may be little more than 1·2 m². Typical office buildings may well have a density of occupation of one person to 10 m² and dwellings may well have a density of occupation of one person to 30-50 m².

2.32 Wider implications

It is clear, therefore, that the occupants at Wallasey contribute a substantial part of the winter heating load. In the summer this heat gain is clearly a disadvantage, but the hope at Wallasey is that the high rates of ventilation which can be achieved by the windows will maintain satisfactory conditions. Tungsten lighting in office buildings is often a problem because of its very high level of heat output and there can be little doubt that at Wallasey the lighting installation was also intended as a heating installation and on cold days is turned on before the pupils arrive in order to give adequate comfort. This is not, of itself, a major criticism. It may well be that if one installation can be used for both lighting and heating the saving in capital cost more than outweighs any increase of energy necessary for thermal comfort. On the other hand it is important that one bears in mind the significance of this aspect in maintaining a thermal balance.

It is clear that Wallasey★ should not be thought of purely as an exercise in solar heating. It is also clear from the density of occupation that there are few other building types which could be designed in this way even if they had appropriate sites. On the other hand, the fact that the school is able to maintain a level of thermal comfort which, if not of a very high order has, nevertheless, been entirely satisfactory to the occupants for a number of years, does give a dramatic demonstration of the importance of taking the form, fabric and fenestration of the building into account when considering the thermal performance.

★ See *The Architects' Journal*, 25 June 1969 for more details.

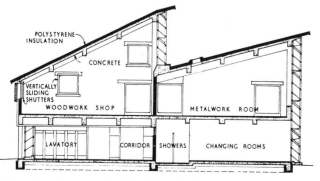

POLYSTYRENE INSULATION

CONCRETE

VERTICALLY SLIDING SHUTTERS

WOODWORK SHOP   METALWORK ROOM

LAVATORY   CORRIDOR   SHOWERS   CHANGING ROOMS

2.5 *section of gymnasium block showing heavyweight construction and positioning of glazing.*

3 ASPECTS OF THERMAL DESIGN

3.1 Comfort and clothing

To consider the thermal performance of a building it is necessary to be familiar with the factors of the physical environment which influence the thermal comfort of the occupants of the building.

Figure **3.2** shows the variables in the human internal thermal environment and table II summarises the factors involved in the building heat balance.

Table II Factors in building heat balance

Initial conditions	+ Heat gains	− Heat losses	→ Satisfactory comfort conditions
Thermal properties of the building. Absolute humidity of the air (unless air conditioning with humidity control is provided).	Solar radiation. Heat from occupants; lighting; mechanical installations and equipment. Heating installation.	Radiation to sky. Convection to air outside. Ventilation losses into ground. Refrigeration plant.	Mean radiant temperature. Air temperature. Relative humidity. (A satisfactory rate of air movement must also be given by ventilation arrangements.)

Note: table II and figures 3.1 and 3.2 are from *Environment and services* by Peter Burberry, Batsford, 1975 (3.1 is based originally on material from BRE CP 61/74).

For thermal comfort to be maintained the human body must lose amounts of heat proportional to the amount of physical activity. This heat can be lost in a number of ways and to a limited degree the balance between the various ways can be varied, such as by choice of clothing.

People can tolerate a much wider variation of environmental conditions when active than when at rest, chart **3.1**. Thermal and thermally related factors which are at work inside buildings must be kept within a relatively narrow range of balance if comfort is to be achieved.

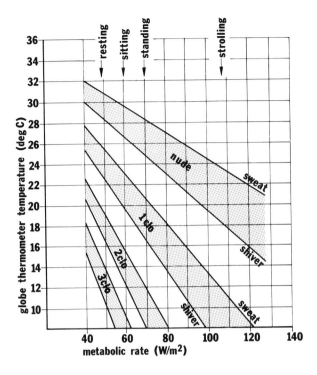

3.1 *Relationship between activity, globe thermometer temperature and comfort.*

3.11 Effect of physical environment on comfort

Development of indices of comfort which could incorporate all these factors into one value for general use in building design have been disappointing. However one simple instrument, the globe thermometer perfected by Vernon in 1930, has been found to give a very close correlation with judgments of thermal comfort; see figure **3.3**

3.12 Effect of clothing on comfort

Though not a part of building, clothing is a factor which cannot be ignored when considering the thermal design of the building. Before the almost universal development of central

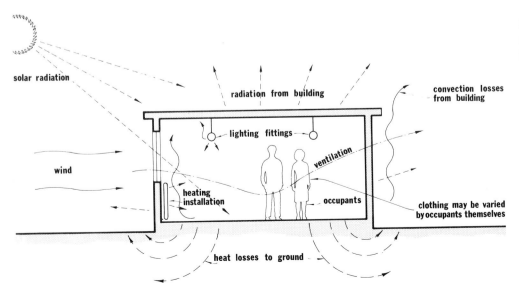

3.2 *Heat balance in relation to internal and external influences.*

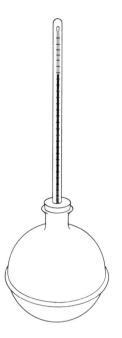

3.3 *Globe thermometer. Bulb of thermometer is at centre of 150 mm diameter matt black sphere.*

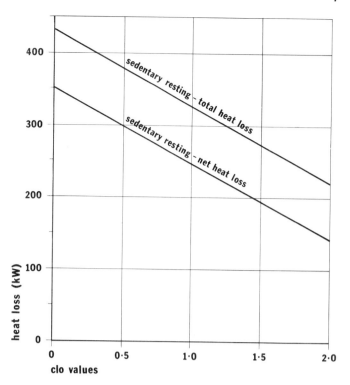

3.4 *Effect of clothing on heat loss takes into account incidental gains from other sources.*

heating the scale of mid-19th century clothing enabled people to sit for services in unheated churches and endure temperatures inside buildings which would now be regarded as intolerable. Present levels of clothing are dramatically reduced though some adjustments are still made to accommodate the body to conditions in buildings.

To make valid thermal predictions and comparisons it was necessary to make some qualification of the thermal effect of clothing. Scales were developed in both England and America; the American scale of units called clo-values is the one which has gained widest acceptance and use. The scale varies from zero for no clothes at all through 1 clo-unit which represents a normal suit and underwear up to a maximum of about 4 which represents heavy polar dress. The unit is scientifically defined in terms of the heat transfer resistance from the skin to the outer surface of the clothed body. Table III shows a typical range of combinations of clothing together with their appropriate clo-value and typical temperatures at which sedentary subjects would be thermally comfortable. It will be observed that comparatively modest variations in clothing have marked effect on comfort temperatures and consequently upon energy conservation in buildings.

3.13 Conclusion

Many people consider that, the present levels of lightness and freedom in clothing having been reached, it would be a very retrograde step to revert to levels of clothing which were common in the past. They suggest that we should maintain our present clothing standards and obtain energy conservation by means of increased insulation or improved plant performance. It would be pleasant if this could be the case; but it is impossible to avoid the conclusion that, as the cost of energy increases, a new balance will be made between amount of clothing and cost of heating which will result in increased clothing levels to some degree. It is certainly the case that if one wishes to make major energy savings in existing buildings without a great deal of capital expenditure, the wearing of more clothing is one of the best ways to achieve this, diagram 3.4.

3.2 Solar energy

3.21 Exploitation by means of solar collectors

Solar energy may be utilised in the overall building design and by means of purpose-made collectors. Collectors placed in the path of the sun's rays enable heat to be transferred to air or

Table III Thermal effect of clothing

clo value		Sedentary and resting Max comfort temp °C
0 clo	Nude	28·5°
0·5 clo	Short underwear Light cotton trousers Short-sleeved, open-neck shirt	25·0°
1·0 clo	Short underwear Typical business suit, including a waistcoat	22·0°
1·5 clo	Long underwear Heavy tweed business suit and waistcoat Woollen socks	18·0°
2·0 clo	Long underwear Heavy tweed suit and waistcoat Woollen socks, heavy shoes Heavy woollen overcoat Gloves, hat	14·5°

3.5 *Milton Keynes house, integral collectors incorporated in roof.*

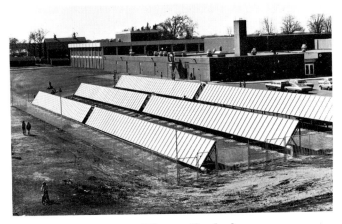

3.6 *Field of collectors at a Minneapolis school.*

water and thence into the building for space or water heating, photographs **3.5**, **3.6**. In many climates the *water heating* application is a particularly attractive one since there is a need for hot water during the summer when solar heat is most readily available. The hot water cylinder which is part of most normal water heating installations provides heat storage capacity; this enables the heat collected during the hottest part of the day to be used at night and even on the subsequent day. The use of *space heating* installations is confined to the winter when, in many climates, solar gains are limited not only by shortness of the day but also by cloud.

Costs

Solar heating installations are expensive; the overall economics have yet to be demonstrated as being fundamentally favourable to solar space heating in the UK. The crucial problem is that of heat storage. During the period when the sun is shining it is likely that proper orientation and fenestration of the building could ensure a useful contribution from the solar heat. For solar space heating installations to play a significant part it is essential that the heat from periods of sunshine can be stored to be used not only overnight, but also during periods of cloudy weather. This requires a substantial volume of storage and substantial expense.

Storage procedures

Gravel and water have been used as media for storing the heat. At present the most promising developments seem to be those involving the use of the latent heat of some salts. The solar heat is used to melt the salt to a liquid condition. During this process the material takes up a large amount of heat in order to achieve its change of state. This heat can then be given up when the salt solidifies. Much work needs to be done in this field, however, before a final clear pattern of utilisation and economics emerges.

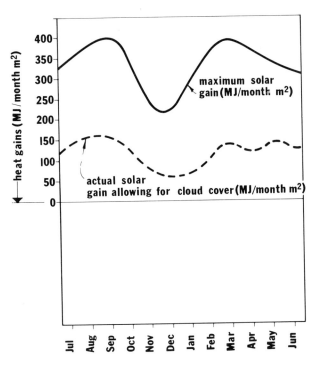

3.7b *Seasonal heat gains through 1 m² unobstructed south-facing glazing.*

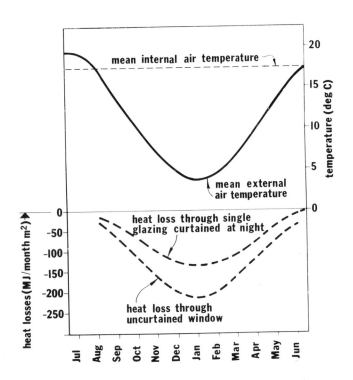

3.7a *Seasonal heat losses through 1 m² single glazing.*

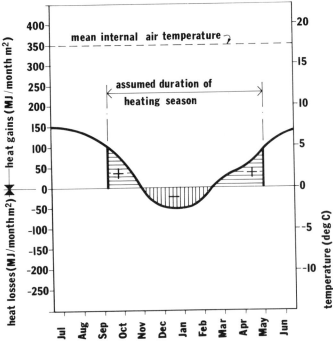

3.7c *Seasonal net heat balance through 1 m² unobstructed south-facing single glazing. (Mean internal air temperature 17·5°C, no curtains.)*

3.22 Exploitation by means of building design

Windows

Alternative means of utilising solar heat lie in the manipulation of form, fabric, and fenestration of the building. Window glass is transparent to the rays of the sun which can pass through it and warm up the interior of the building. The glass is opaque to the radiation coming from the surfaces inside the building, diagram **3.9**; windows do, however, lose heat as a result of normal convection. If the radiation gain can exceed the convection loss then windows will contribute heat to the interior of the building. Where windows are completely uncurtained by day and night the convection losses exceed the radiation gains. However on south facing facades, if windows are curtained at night, the solar radiation gained during the heating season will exceed the losses through the window.

Orientation

The results are shown in diagrams **3.7a** to **3.8f** of an analysis of the radiation gains and the convection losses through 1 m² of window glass with no serious obstructions to cut off the sun. For **3.7** a calculation is made of mean flow rate for each month. For solar radiation **3.7b** shows the amount which would pass through the window with a completely clear sky and a second curve, more relevant to design, corrects the original one to take into account typical cloud cover. The convective loss curves for 24 hours, **3.8a–f**, show the heat loss through the window resulting from an internal mean temperature of 20°C, **3.8a, c**, and a second curve, again more relevant to design, shows the modification of this loss as a result of drawing reasonably substantial curtains at night, **3.8b, d**. The charts show clearly that for the south facing window, unobstructed and curtained at night, there is an overall heat gain during the heating season. This is not the case on other orientations but there is nevertheless substantial variation in the amount of heat loss.

It does not follow that if buildings were designed with all their windows facing south the heating installation would be unnecessary. It would be necessary to provide heat for overcast periods and the larger the window, the larger the size of installation required. On the other hand, it is quite clear that for windows of desired size, careful attention to planning the building in order to give the best possible orientation would have a significant return in energy saving.

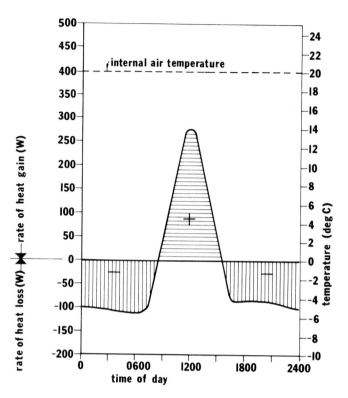

3.8b *Net heat balance through 1 m² unobstructed south-facing glazing at mid-January*
1 *Single glazing*
  Internal air temperature 20°C
  No curtains.

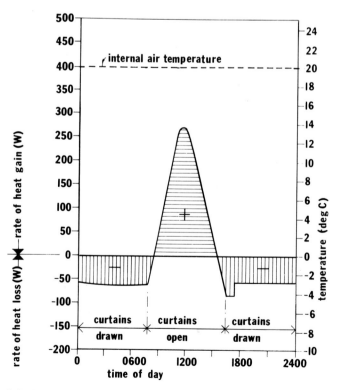

3.8c *Net heat balance through 1 m² unobstructed south-facing glazing at mid-January*
2 *Single glazing*
  Internal air temperature 20°C
  Curtains drawn at night.

3.8a *Typical pattern of gross heat gains and losses through 1 m² unobstructed, south-facing single glazing at mid-January.*

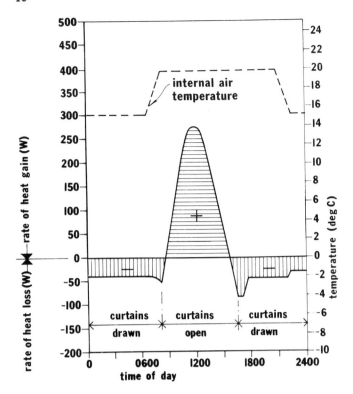

3.8d *Net heat balance through 1 m² unobstructed south-facing glazing at mid-January*
3 *Single glazing*
Internal temperature 20°C reduced at night
Curtains drawn at night.

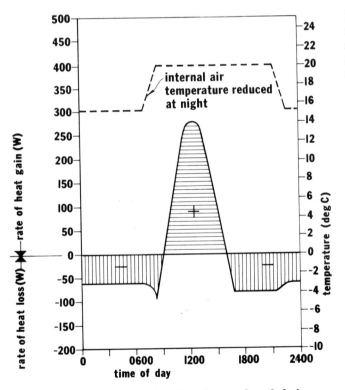

3.8e *Net heat balance through 1 m² unobstructed south-facing glazing at mid-January*
4 *Single glazing*
Internal temperature 20°C reduced at night
No curtains.

Internal finishes

The degree to which this solar heat gain can be utilised depends a great deal upon the arrangement of the internal finishes. If the sun's rays, after penetrating the window, fall upon carpet and lightweight materials the heat will immediately be liberated into the air; the room will overheat and will have to be kept cool by excessively high levels of ventilation.

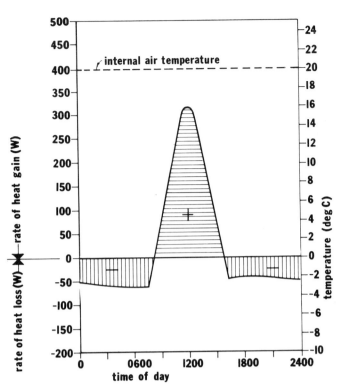

3.8f *Net heat balance through 1 m² unobstructed south-facing glazing at mid-January*
5 *Double glazing*
Internal temperature 20°C
No curtains.

If on the other hand the sun's rays fall upon fairly massive materials having substantial capacity for heat storage, then much of the heat will be absorbed into the structure, thereby reducing the initial surge in temperature but contributing heat to the environment after the direct rays of the sun have gone, see diagram **3.10**.

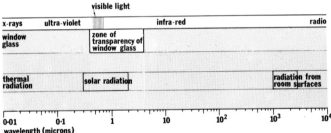

3.9 *The electro-magnetic spectrum showing the wavelength range of window transparency compared with the wavelength of thermal radiation. Most solar radiation will pass through windows but internal radiation, of a changed wavelength, finds windows opaque at that wavelength.*

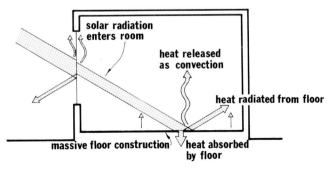

3.10 *Massive floor construction absorbs heat rapidly. Lightweight or carpeted floors will reach high surface temperatures and a large proportion of the heat will immediately be liberated into the room giving higher peak temperatures.*

This aspect of design can only be applied to new buildings, where form and orientation can be controlled, but it is an extremely attractive one from the energy conservation point of view. It is possible to produce buildings which cost no more than they otherwise would and yet which make a small but significant saving in energy.

Design possibilities

It would clearly be possible to adapt the design of buildings beyond the present conventional limits in order to take advantage of the full potential of solar gain. The general principles which would be appropriate are clear from the preceding paragraphs. Lightweight finishes which would convert the solar rays to convected heat quickly should be avoided. Massive elements of construction should be placed where the sun's rays can warm them. Orientation and window size should be carefully considered together with screening and shuttering of windows to give reduced heat loss at night and to control extremes of summer overheating. These principles of providing solar warming were exemplified by Xenophon some 2000 years ago.

In most design at present we have not really reached the level which he postulated. Window shutters have completely disappeared from modern design. Any consideration of thermal problems, however, leads us to the conclusion that shutters are potentially excellent features. They enable windows to be provided with a high level of insulation at night, together with substantially reduced ventilation rates. They can be opened up to allow penetration of light, solar heat and air. Many modern designs observed abroad also provide canopies which enable the sun's penetration to be controlled in very hot weather. There is an urgent need for development of shutters and their deployment in buildings.

3.3 Thermal behaviour of buildings

3.31 'Steady state' concept

The normal convention both for thinking about the thermal properties of buildings and for estimation of plant size has been the concept of 'steady state' conditions.

An estimate is made of the rate at which heat would be lost from the building assuming the outside and inside temperatures remain constant over long periods. The figure resulting from this calculation, adjusted to take into account boiler efficiency, allowance for warming up etc, indicates the plant size required. Modifications of the calculation give an idea of seasonal heat requirements.

Changes in building fabric

For many years buildings were reasonably massive with small windows and heating was mainly by means of low pressure hot water heated by solid fuel boilers. With this type of building and the slow response type of heating installation the steady state method of analysis gave very acceptable results. For a substantial time, however, the thermal properties of buildings have tended to change towards lighter weight which has a lower thermal capacity and consequently gives rise to much greater temperature fluctuations for a given heat input. New types of fuel, increased acceleration in the distribution system, improved emitters and much improved automatic control systems contribute to heating installations which are able to react very much more rapidly to changing circumstances.

Changes in building use

New patterns of usage exist, particularly in the domestic field, where intermittent occupancy of buildings calls for rapid warming up. It has become necessary to take a very much more sophisticated view of the way in which buildings behave thermally in order to relate the type of fabric to the type of installation and to relate both of these to the use required. It is necessary not only to take into account the insulation value of the materials of which the building is made, but also the capacity of these materials to absorb heat.

Intermittent use

If a building is to have intermittent use and it is desired to take advantage of the possibilities which this offers for fuel saving by turning off the heating, it would be most inappropriate to use a very massive form of construction: the heating installation would have to be turned on many hours, if not days, before the next use in order to warm up the fabric of the building and ensure that the mean radiant temperature from the walls was high enough.

Continuous use

Similarly if intermittent usage is not required and a massive form of construction has been selected there is probably little point in selecting an installation with extremely sophisticated controls and rapid response since these will serve little useful purpose. It is not always necessary to select a complete lightweight construction in order to achieve rapid response to thermal conditions inside a building: an appropriate inner lining of low thermal capacity can transform a room from slow to quick response to heat input.

3.32 Dynamic behaviour of buildings

Table IV shows the thermal capacity of walls of different materials having the same U-value. The difference in thermal

Table IV Variations in thermal capacity for constant U-value

Material	Density kg/m³	Specific heat J/kg °C	Volumetric specific heat kJ/m³degC	Conductivity 'k' W/m °C	Thickness for U of 1·0 W/m²degC mm	MJ required to raise this thickness for U of 1·0 by 20°C	Temperature rise resulting from application of 1 kW for 1 minute °C
Concrete	2100	840	1760	1·0	830	29·2	0·04
Brickwork	1700	800	1360	0·84	700	19·0	0·06
Timber	600	1210	730	0·14	120	1·7	0·68
Lightweight concrete	1000	1000	1000	0·3	250	5·0	0·24
Wood wool	500	1000	500	0·1	83	0·83	1·4
Fibreboard	300	1000	300	0·05	42	0·25	4·8
Expanded polystyrene	25	1000	25	0·03	25	0·01	96·0

Note: a joule is the SI unit of work, energy and heat (equal to 0·239 calories). kJ is 10³ joules, MJ is 10⁶ joules, GJ is 10⁹ joules.

capacity is dramatically apparent. Upon this depends the speed of response of the building to fluctuations in heat input. The figures for warming up periods shown in the table demonstrate very effectively the differences in the time that would be taken to reach comfort conditions in rooms formed from the materials specified.

Continuous use

In buildings which are continuously heated there will be very little difference between the heat losses of different constructions having the same U-value. Some marginal differences may be discernible even here since high capacity structures lessen the effect of control systems. For practical purposes there will be little significant difference. However, in reality very few buildings are maintained at a fixed temperature day and night. Prisons and hospitals are the best known types which do have this pattern of heating.

Intermittent use

Almost all other buildings have their heating either turned off or operating at reduced temperatures at night, and in many commercial and office buildings at the weekend also. At night and at the weekend all these buildings will cool down, losing heat not only through their fabric but also by ventilation. They will then have to be warmed to comfortable temperatures to be ready for occupation.

Examples of dynamic heat flow

The temperature distribution and direction, and magnitudes of heat flow in both steady and dynamic conditions for three types of wall, are shown in diagrams 3.11 to 3.13. 3.11 **a** to **f** show a homogeneous wall; 3.12 **a** to **f** a wall with insulation on the inner face; and 3.13 **a** to **f** a wall with a cavity or with insulation at its centre.

3.33 Lightweight and heavyweight buildings

Heavyweight buildings of high thermal capacity will not, because of the amount of heat stored, cool to such a low temperature as lightweight buildings; but in similar conditions they will lose more heat than the lightweight construction and will require more heat input than a lower capacity building to bring them back to comfort level.

Table V demonstrates the nature of this phenomenon. Many lay observers find it difficult to understand that it can take more heat to raise a building which is warmer to comfort temperature than one which started at a significant lower temperature. It is a concept, however, which it is essential to grasp when considering the thermal behaviour of buildings.

Modifying heavyweight fabric performance

It is usual to equate heavyweight buildings with high thermal capacity and broadly this is the case. It is, however, possible to influence the effective capacity of the building by means of insulation and finishes. A massive construction of concrete, if carpeted and provided with a suspended ceiling of insulation board and internal insulation on the walls will, so far as indoor temperatures are concerned, behave in the same way as a lightweight construction.

3.34 Fabric performance and fuel consumption

The graph 3.14, based on data given in the IHVE guide, shows the effect of thermal capacity on fuel consumption together with the effect of the response available from the plant. For a plant capable of heating the building on a continuous basis (plant size ratio 1) there is little saving from intermittent use

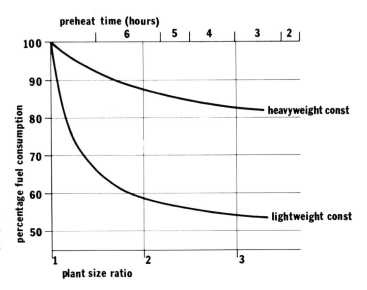

3.14 *Relationship between fuel savings and thermal design, comparing the benefits of lightweight and heavyweight construction. Based on building occupied 8 hours/day, 7 days/week. (Plant size ratio of 2 means twice plant size.)*

Table V Heat required to raise fabric temperature

Material	Assumed drop in temperature °C	Heat required to return to required ambient temperature kJ/m²
100 mm brickwork	5	880
100 mm fibreboard	20	600

Note: the relatively low thermal capacity material, fibreboard, will cool to a much lower temperature (by 20°C) in the same period than brick (by only 5°C) which has a relatively higher thermal capacity. But the fibreboard (low capacity) will require less heat to restore its original temperature (reheating through 20°C) than brick (high capacity, reheating through 5°C). The fibreboard lined room would, of course, be colder during the cycle.

when compared with continuous heating (100 per cent fuel consumption). However, if the capacity of the plant is increased so that quick warming up becomes possible (plant size ratio greater than 1) there is a dramatic saving of fuel. With a heavyweight building the saving is limited to about 85 per cent of the continuous requirement but with lightweight construction the fuel requirement can drop to 56 per cent.

Design implications

It is evident therefore that for economy of energy use during the heating season a low thermal capacity construction is best. Unfortunately the reverse is true of summer conditions. High thermal capacity walls and floors will absorb heat in summer, particularly when the direct rays of the sun impinge on them. Lightweight construction will warm up rapidly, giving higher peak temperatures. Much can be done to control summer temperatures by means of carefully designed fenestration and adequate ventilation. In this climate, therefore, where in normal buildings conditions need not become intolerable in summer, it is likely that economy of energy in winter will govern the choice of the internal capacity of the fabric. It appears, however, that no studies directed towards establishing optimum solutions have been undertaken. This should be rectified urgently.

3.35 Steady state and dynamic concepts related to heat loss

Use of analogue prediction methods

Traditionally the only thermal estimates made were of the

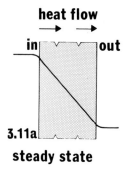

heat flow

in | out

3.11a

steady state

3.11 to 3.13 Low temperature variations across wall. Arrows indicate direction of heat flow. Dimensions marked with an asterisk indicate temperature difference between internal wall surface and internal air. This difference considerably influences comfort and condensation.

Homogeneous wall (**3.11a** to **f**)

In the steady state a continuous flow of heat takes place from inside to outside. In dynamic conditions heat flows back into the room as the internal temperature drops; the temperature of the wall also drops and the wall will require heat to warm it next day when heating begins.

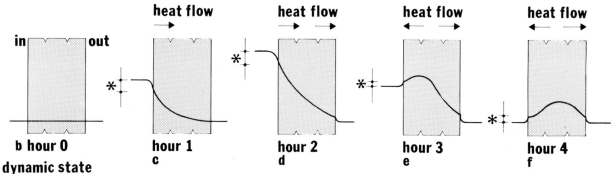

b hour 0
dynamic state
    hour 1
    c
    hour 2
    d
    hour 3
    e
    hour 4
    f

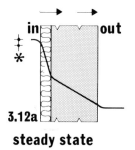

3.12a

steady state

Insulation on inner face (**3.12a** to **f**)

Substantially less heat is absorbed by the wall. For a given heat input the internal air temperature will rise more quickly since less heat is passing into the wall. Comfort conditions will be improved since the inner surface temperature is much closer to air temperature during the warming up period.

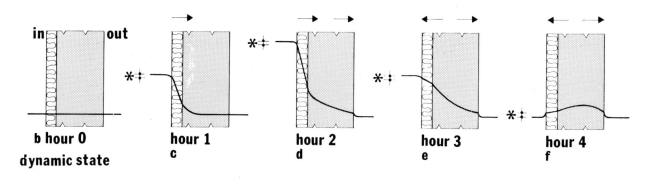

b hour 0
dynamic state
    hour 1
    c
    hour 2
    d
    hour 3
    e
    hour 4
    f

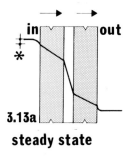

3.13a

steady state

Air or other cavity insulant (**3.13a** to **f**)

The same phenomenon with performance intermediate between a homogeneous wall and insulation on inner face.

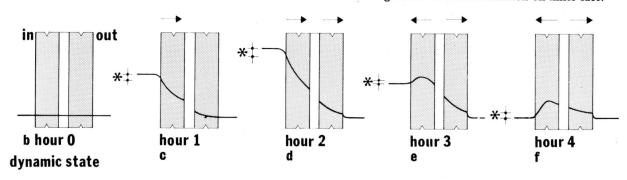

b hour 0
dynamic state
    hour 1
    c
    hour 2
    d
    hour 3
    e
    hour 4
    f

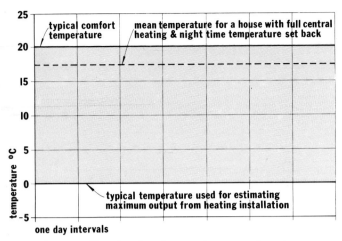

3.15a *Concept of heat loss based on steady state and maximum design temperature difference. Tone represents heat loss.*

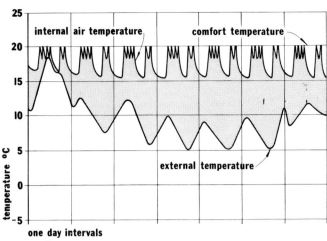

3.16a *Typical internal and external temperature variations in dwellings with central heating. Compare with* **3.16b**.

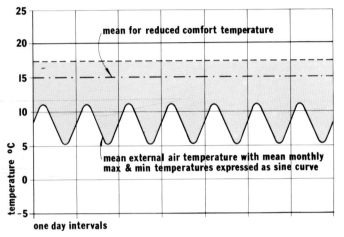

3.15b *Concept of heat loss based on internal and external mean temperatures. Tone represents heat loss. Note the substantial reduction as a result of reducing comfort temperature.*

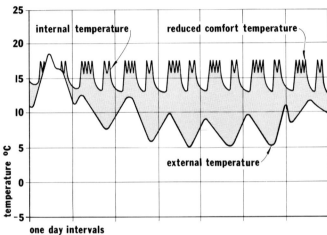

3.16b *Typical external air temperature and reduced internal comfort temperature (from 20°C to 17·5°C). Compare with* **3.15a, b** *and* **3.16a**.

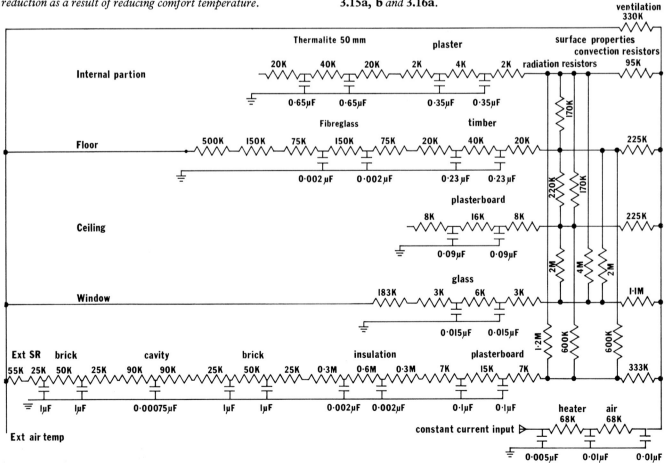

3.17 *Electrical representation of thermal properties used for simulating the conditions and achieving the results shown above.*

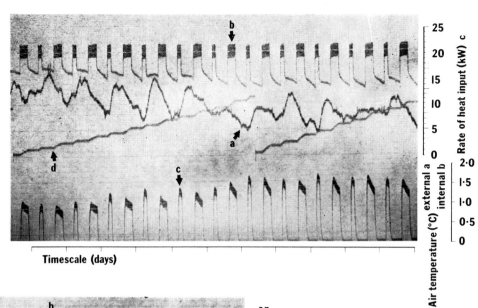

3.18 *Sections of output charts from the thermal analogue. In each case the external air temperature is recorded (**a**), together with the internal air temperature (**b**), the rate of heat input (**c**), and the cumulative heat input (**d**). The deep band on trace (**b**) shows the small air temperature variation resulting from the operation of the thermostat. The cumulative heat input trace (**d**) is returned to zero at the end of each week and the seasonal total calculated as the sum of the weekly totals.*

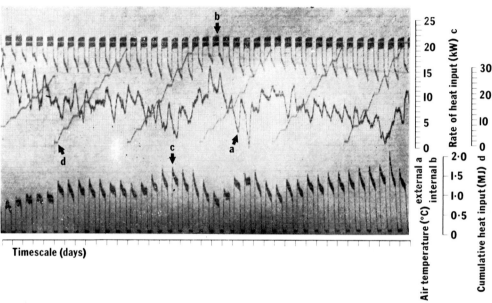

maximum rate of heat loss. It is only recently that energy consumption has been closely considered. There is still a tendency to think in unmodified maximum rate terms.

Graph **3.15a** shows the consequences of this. Savings due to reduced temperatures would be minimised while savings due to insulation would be exaggerated.

Graph **3.15b** shows a more realistic pattern based upon mean external and internal temperatures, taking into account daily and seasonal variations. It is apparent that a reduction in internal comfort temperature of 2 or 2·5°C will give a very significant reduction in heat loss, of the order of 25 per cent.

Table III, giving the relationship between levels of clothing and comfort temperatures, demonstrated that a reduction of temperature of this degree could be achieved without dramatic changes in clothing.

Graphs **3.16a, b** show, for part of a heating season, comparisons of actual internal and external conditions for comfort temperatures of 20°C and 17·5°C.

Results of analogue studies of insulation position and heating pattern

The analyses described above use steady state techniques (where fixed internal and external conditions are assumed for calculation purposes) and could in practice be influenced by the position of the insulation and the pattern of use of the building. For continuous heating the position of the insulation would not be critical and an estimate of the seasonal consumption could be based upon the rate of heat loss. In time of energy shortage, however, continuous heating is to be avoided. Most domestic buildings have intermittent heating and the situation is potentially very different. An insulating material

on the inner face of the wall would warm up quickly when heat is input to the room and thereby give potential economy provided the heating installation can respond to control. If, as seems to be gaining in popularity, the insulation is placed in the cavity, then the inner leaf will require more heat to bring it to a satisfactory temperature, and during the night it will lose the heat and require to be warmed again next day, thus increasing the energy requirement.

It is not normally possible to make analyses which will demonstrate the effect of this phenomenon over a whole heating season. It is clearly very desirable to do so, however, and a thermal analogue has been applied to this study. Tables VI and VII show the results. The tables indicate that, for intermittent heating, the thickness of insulation is far less important than its position. As might be expected the difference is much less significant in continuous heating but this can hardly be regarded as an acceptable system when the very substantial differences between it and intermittent heating are studied. In order to make this dynamic analysis, an electrical model was made of the thermal properties of a room in a terrace of housing. A typical configuration is seen in diagram **3.17**. Voltages and currents represent external temperatures and heat inputs were applied to the model. The heat inputs were electronically controlled to represent a time and temperature controlled heating system. Several different time regimes for heat input were adopted and the internal air temperature controlled at 20°C. The simulation was carried out over a period of time representing a heating season of 31 weeks using Meterological Office data on weather at Kew. Internal temperatures reached were recorded continuously together with the rate of energy input and the cumulative total of energy supplied. Some of the output used in achieving the results shown in tables VI and VII is illustrated in output charts **3.18**. The technique is one which clearly has important application to problems of energy saving.

Lessons of analogue studies of insulation position and heating pattern

Tables VI and VII and diagram **3.19** express in different forms the results of the analogue studies. They demonstrate the vital importance of time control of heating installations. Other than hospitals and prisons few buildings require continuous heating and a very substantial proportion of dwellings are unoccupied for much of the day. Both tables and chart demonstrate that for any degree of insulation dramatic reduction in energy consumption can be achieved by intermittent operation of the heating to whatever degree the occupancy of the building makes possible.

25 mm of additional insulation to the brick cavity wall has a significant effect on reducing heat loss. Further thickenings have a diminishing effect. In the case of continuous heating the position of the insulation does not affect the heat loss significantly. Where intermittent heating is employed, however, better economy of energy is achieved when the insulation is on the inner face of the wall rather than in the cavity. This is particularly the case with fast responding heating systems such as the electric fan convector. The slower responding hot water system reduces slightly the significance of locating the insulation on the inner face.

These results are based on a room with only one external wall. While the principles would remain the same the scale of saving would be greater for rooms with more than one external wall and for complete buildings.

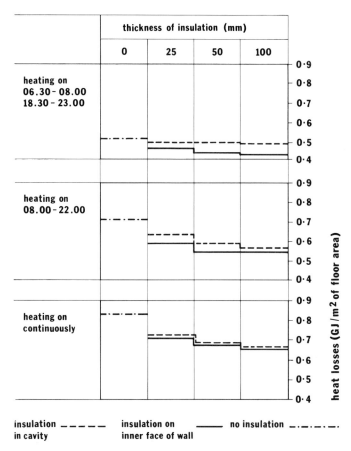

3.19 *Energy consumption with variation of heating pattern and thickness and position of insulation (2kW electric fan convector).*

Table VI Comparative energy input during heating season in GJ/m² of floor area*

(Based on 20 m² room in terrace housing)
This table and table V show the greatly increased energy input due to continuous heating as against intermittent heating

Heating regime	None	In cavity			Inner face of wall		
		25 mm	50 mm	100 mm	25 mm	50 mm	100 mm
2 kW electric fan convector							
2 period†	0·512	0·494	0·491	0·487	0·458	0·431	0·426
14 hour on	0·713	0·632	0·584	0·566	0·589	0·542	0·541
Continuous	0·840	0·728	0·680	0·665	0·719	0·679	0·654
2 kW water/natural draft convector							
2 period	0·565	—	0·510	—	—	0·482	—
3 kW electric fan convector							
2 period	0·571	—	—	—	—	—	0·436

* Gaps indicate that research work has not yet been completed.
† 6·30 to 8·00 and 18·30 to 23·00.

Table VII Comparative energy input during heating season shown as percentage related to 2 kW, 2 period (6·30 to 8·00 and 18·30 to 23·00), 100 mm insulation on inner face as 100 per cent

(Based on 20 m² room in terrace housing)

Heating regime	None	In cavity			Inner face of wall		
		25 mm	50 mm	100 mm	25 mm	50 mm	100 mm
2 kW electric fan convector							
2 period	120*	116*	115*	114	108	101	100
14 hours on	167	148	137	133	138	127	127
Continuous	197	171	160	156	169	159	154
2 kW water/natural draft convector							
2 period	132*	—	119*	—	—	113	—
3 kW fan convector							
2 period	134*	—	—	—	—	—	102

* For a few days in the cold weather comfort is not maintained.

4 SIGNIFICANCE OF THERMAL FACTORS

4.1 Main design variables

Relative importance of the thermal factors controlled by decisions about the building

It is impossible, and inappropriate, to give specific design values to the importance of the various ways in which building design can affect heat loss. Their relative significance will vary from building to building, and even in the same building one decision may alter the significance of other factors (eg a decision to use lower internal temperatures reduces the economic effectiveness of insulation). It is, however, very important for designers to be aware of the factors involved and the approximate order of their significance in typical cases.

The following paragraphs give information of this sort for the main design variables:

4.11 Siting
4.12 Multiple use
4.13 Volume
4.14 Shape
4.15 Grouping
4.16 Internal planning
4.17 Insulation
4.18 Ventilation
4.19 Thermal response
4.20 Fenestration and orientation
4.21 Clothing.

The notes below summarise the design decisions and their general quantitative significance.

4.11 Siting

The effect of moving the notional house forms used in **4.1**, **4.2** (illustrated in key diagram **4.1**) from a sheltered site to one with severe exposure and increasing the ventilation rate by 25 per cent to take account of wind effects is to increase the overall rate of heat loss by 20 per cent. However, designers seldom have a choice of site and this consideration is one for town planners. While increase in population and shortage of land clearly have an important influence on land selected for building, local knowledge of the exposure of sites which had an important influence, has in recent years been ignored.

4.12 Multiple use

If two uses can be accommodated in one building, virtually the whole of the energy required for one will be saved. If the two combined uses mean that some, but not all, of the separate accommodation which would be required must be built, the savings are reduced but will still be very great.

4.13 Volume

The variation of heat loss with increase in volume is shown alongside. How the heat loss varies with volume is shown in **4.2** while **4.3** takes into account the typical heat gain from a small

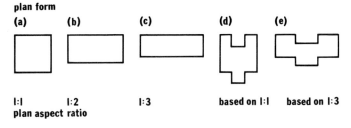

plan form

(a)   (b)   (c)   (d)   (e)

1:1   1:2   1:3   based on 1:1   based on 1:3
plan aspect ratio

4.1 *Various plan forms used as basis for charts 4.2, 4.3*

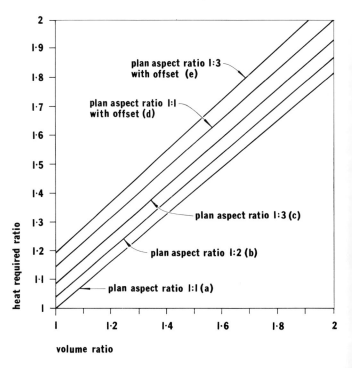

4.2 *Curves show how the ratio of heat loss varies with volume for various plan forms. (Based on 100 m² two storey dwelling with square plan form as unity. Volume ratio here assumes 2 storey and constant storey height. Thus volume ratio of say 2 implies 200 m² floor area on two floors. Dwelling with heat loss ratio of say 1·5 has heat loss 1·5 times that of basic, 2 storey 100 m², dwelling).*

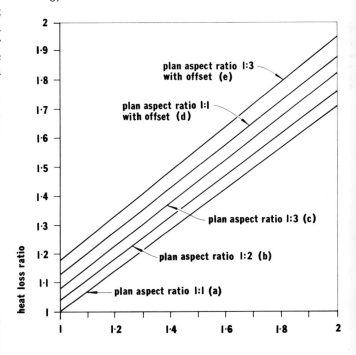

4.3 *Curves show how the ratio of heat required varies with volume for various plan forms. (Takes into account 20GJ per season heat gain from occupants, cooking, water heating, etc. Solar gain is not included. Dwelling with heat required ratio of say 1·5 has heat required 1·5 times that of basic, 2 storey 100 m², dwelling.)*

family. It will be observed that the increase in heat requirement is almost directly proportional to the increase in volume and that economical design in terms of reducing space in the building yields major dividends in energy saving.

4.14 Shape

The variation of heat loss resulting from considerable variations of shape are shown in **4.2**, **4.3**. Moving from the economical square plan shape (a) to the very different 1:3 aspect ratio with offset (e) results in an increase in heat loss of less than 20 per cent. It appears, therefore, that variation in shape within these limits is not a major consideration in heat loss.

More extreme variations of shape, such as might result from transforming an aspect ratio of 1:1 into an aspect ratio of 1:10, must be rare and it is difficult to think of functions which can be adequately met by both plan forms without major variations in circulation area. This would mean in the case of the elongated form that the area of the external surface and the consequent heat loss would be substantially increased. For similar windows, the ventilation rate would also be increased for most buildings. In addition, if the volume of the building were increased by additional circulation the heat loss would be proportionally increased.

Extreme forms of these aspect ratios are often found in multi-storey buildings and here the heat losses are likely to be increased by the additional exposure of the high building and the energy costs possibly increased by the need for air conditioning (because the height of the building and the lack of screening from trees and other buildings result in stack effects and solar gains necessitating air conditioning).

4.15 Grouping

While variation of shape within practical limits has little effect upon heat loss, separation of blocks, or conversely combining separate buildings into larger blocks has a major effect on energy requirements. Chart **4.4** shows the variation possible with different typical house arrangements. Flats would, of course, demonstrate an even more marked economy of energy.

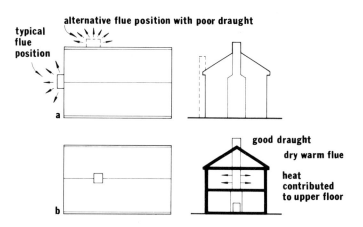

*4.5 Bad and good flue positions. In **a** the external flue is relatively wet and cold which decreases the draught and increases condensation.*

4.16 Internal planning

Location of heat sources

In dwellings the major item of heat emitting plant which can be located to give economy of energy is the flue. In a central position the flue contributes heat to upper floors while on an external wall it wastes heat, is cold, is often badly placed for draught and liable to condensation. In a dwelling with central heating up to 5 per cent of fuel may be saved by appropriate flue position. Diagram **4.5** shows the considerations involved.

Open plan

No generalised information can be provided but the consequences of this type of planning in terms of higher average building temperature are apparent.

4.17 Insulation

In the days when a single open fire sufficed for heating a house (ie most of the house was not heated; and draughts and open flues were commonplace) insulation would have served no useful purpose and formed no part of the British nineteenth century house. With the development of heating, higher levels of insulation became important, not so much to save energy as to provide comfort by ensuring an adequate surface temperature for walls and ceilings. The present boom in the sale of double glazing perhaps typifies the situation. It is very doubtful whether double glazing will save money, even at present fuel prices, although it can give improved comfort.

Insulation, in the present situation, does seem to present an obvious way of saving energy. In a badly insulated structure the effect of a layer of insulation will be dramatic. However, the effectiveness of adding further layers will rapidly diminish. The relationship between U-value (the heat transmittance of a construction) and rate of heat loss is shown in graph **4.6**. This is the basis for the subsequent charts dealing with specific instances. This chart covers the total range of possible heat loss and U-values, from the ultimate minimum when the U-value of 0 represents an infinite insulation, to a U-value of 5 which results from sheltered external and internal surface resistances only with no additional contribution from the building material itself. If an actual U-value for a particular wall is selected along the bottom scale, projected upwards to cut the vertical line and then horizontally, the appropriate rate of heat loss for a 20°C temperature difference can be read immediately. Chart **4.7** shows a series of such lines representing the rates of heat loss for a single 9·5 mm asbestos cement sheet and for the progressive addition of 12 mm layers of insulation. The chart shows very clearly the dramatic reduction of heat loss achieved by the first layer and the rapidly reducing savings resulting from subsequent layers.

Internal surface temperature can be an important factor in

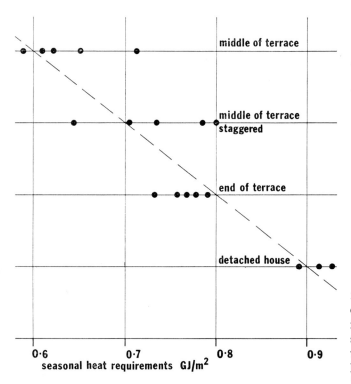

0·6   0·7   0·8   0·9
seasonal heat requirements GJ/m²

4.4 Analysis of comparative heat requirements for typical terraced and detached houses. The dots indicate specific results and the dotted line shows the increase in heat requirements from enclosed terrace houses to isolated detached dwellings. Unit of measurement is the joule, the unit normally used for energy. GJ, or giga joules, is 10⁹J.

thermal comfort and is vitally affected by insulation. Chart **4.8** shows the basic relationship between U-value and inner surface temperature for walls for 20²C temperature difference. Chart **4.9** shows the variation in surface temperature resulting from the progressive layers of insulation on an asbestos cement sheet. The rapid reduction in effectiveness with increasing thickness of insulation is immediately obvious.

It must be remembered that the cost of insulation follows the same pattern. Additional thickness installed at the same time is relatively less expensive than an original layer.

It may be thought that insulation applied to an asbestos cement sheet is a special case and that conventional wall construction would behave differently. Charts **4.10** and **4.11** show precisely the same effect when insulation is applied to a brick wall. But note that the effectiveness of the first layer of insulation itself is much *less* than in the case of the much less resistant asbestos cement.

The effectiveness of insulation is also governed by its location. This point is dealt with in paragraph **3.35** above.

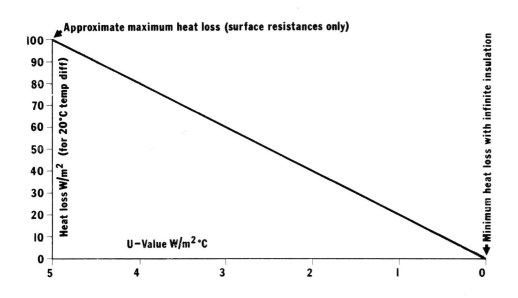

4.6 *Basic relationship between U-value and rate of heat loss. The figures at the extreme ends of the scales are the theoretical maximum and minimum values possible. This graph is the basis for* **4.7** *and* **4.10**.

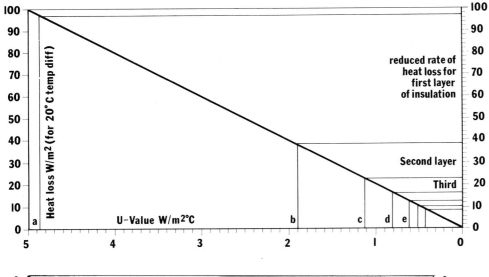

4.7 *Comparative effectiveness on the rate of heat loss of successive layers of glasswool insulation applied to 9·5 mm asbestos cement sheeting with a 20°C external/internal temperature difference. The first insulating layer produces startling reduction in heat loss, but the benefits of insulation are considerably diminished as each successive layer is applied; The U-values are shown as follows:* **a**, *basic loss through sheet,* **b**, *with 12 mm glasswool,* **c**, *with 24 mm,* **d** *with 36 mm,* **e** *with 48 mm.*

4.8 *Relationship between U-value and inner surface temperatures for normal exposure and with a 20°C internal/external temperature difference. The figures at the extreme ends of scales are the theoretical maximum and minimum design values. This graph is the basis for* **4.9** *and* **4.10**.

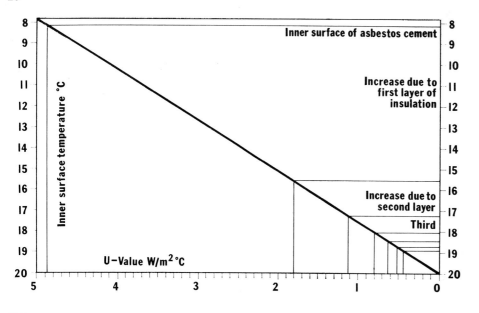

8 Inner surface of asbestos cement

Increase due to first layer of insulation

Increase due to second layer

Third

Inner surface temperature °C

U−Value W/m²°C

4.9 *Comparison of inner surface temperatures achieved by applying successive layers of glasswool insulation to 9·5 mm asbestos cement sheet, with a 20°C internal/external temperature difference. As in* **4.7** *the immediate effect is dramatic, but successive layers give increasingly less benefit.*

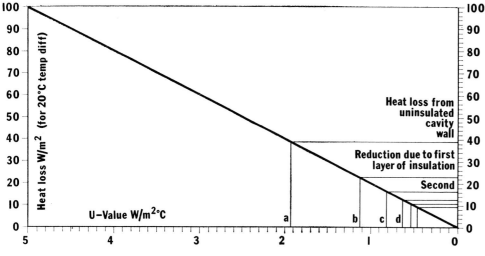

Heat loss from uninsulated cavity wall

Reduction due to first layer of insulation

Second

Heat loss W/m² (for 20°C temp diff)

U−Value W/m²°C

a   b c d

4.10 *Comparative effectiveness of successive layers of foamed polystyrene insulation on rate of heat loss of a brick cavity wall, with a 20°C internal/external temperature difference. The U-values are shown as follows:* **a** *uninsulated cavity wall,* **b** *with 12 mm polystyrene,* **c** *with 24 mm,* **d** *with 36 mm. Again the same immediate (but less) benefit as in 8, followed by a sharp reduction in effectiveness. See also* **4.11**.

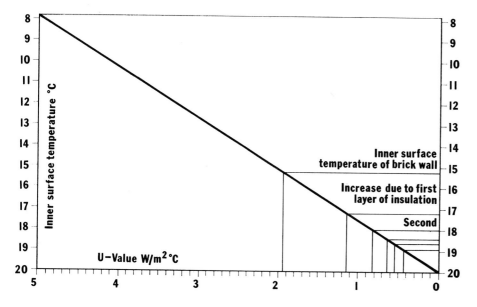

Inner surface temperature of brick wall

Increase due to first layer of insulation

Second

Inner surface temperature °C

U−Value W/m²°C

4.11 *Same construction and insulation as in* **4.10** *and comparing effectiveness of successive layers of insulation on internal surface temperatures of a brick cavity wall.*

4.18 Ventilation

Ventilation is a major reason for winter heat losses and many of the public pronouncements on energy have called for reduced levels of ventilation. Graph **4.12** shows how, for a typical commercial building, heat loss varies with ventilation rate. Much could be done to reduce heat losses by reducing excessive ventilation.

There is, however, as has been shown by the great increase in condensation problems since flues or permanent vents were no longer legally required in houses, a considerable risk of creating condensation problems by the restriction of ventilation. In building design there have been too many fundamental errors made by concentrating attention upon only one aspect, and making changes without considering their consequences. This situation must not be repeated again.

Before any attempt is made to achieve substantial reductions in ventilation rates careful predictions and practical tests must be made.

4.19 Thermal response

See paragraphs **3.31** to **3.35** above.

4.20 Fenestration and orientation

The general principles involved have been covered under Solar Energy, paragraphs **3.21** and **3.22**. The following figure summarises the net heat losses for various orientations of 1 m² of unobstructed window. It is apparent that efforts should be made in design to achieve the best possible orientations for windows.

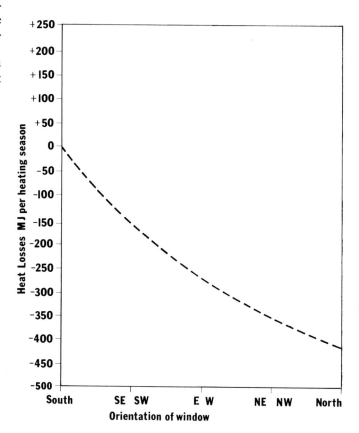

4.13 *Net heat balance of* 1 *m*² *unobstructed clear glass window, curtained at night with* 17·5°C *mean internal temperature, for various orientations.*

A south facing window shows a gain during the heating season. It does not follow, however, that south facing windows should be increased in size for this reason. There are substantial periods without sunshine and additional plant would be required to maintain comfort during these periods. Larger windows may also contribute to overheating in summer.

4.21 Clothing

Although not strictly a part of building, clothing is a factor that cannot be ignored. Before central heating much of the balance in thermal environment was made by means of clothes. In order to make valid thermal predictions and comparisons some quantification has had to be made of the thermal effects of typical clothing, resulting in a scale of clo values which can be related to comfort at various temperatures (table III, p. 7). The rates of heat loss of varying clo values are shown in **3.4**. Comparatively modest variations in clothing have marked effects on energy conservation in buildings.

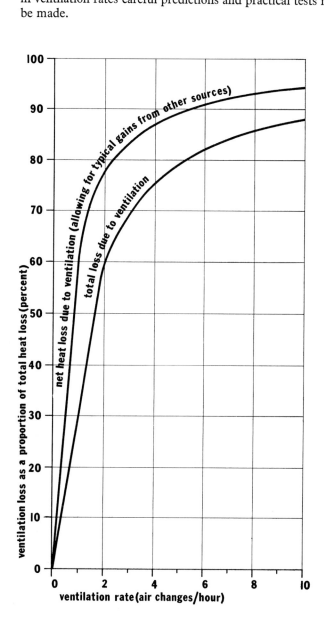

4.12 *Proportion of heat lost through ventilation in typical office building.*

5 THERMAL REGULATIONS

5.1 Purpose of legislation
5.11 Legislation and safety

Legislation controlling the design and construction of buildings has its origins, in most countries, in considerations of safety and particularly fire precautions. The inclusion of requirements for the thermal performance of buildings is of very recent origin. In England and Wales although individual local authorities imposed insulation standards, there were no overall official requirements for thermal performance until 1953 and the current standards in many other countries date from this time and later.

It is not appropriate to schedule present legislation in detail. Requirements in this field are changing rapidly and in any actual design it is essential to consider the current standards. It is, however, appropriate to describe and exemplify some of the development of legislation for controlling the thermal performance of buildings and to consider what legislation might be appropriate for energy conservation.

5.12 Rate of heat transmission

Most present legislation controls the maximum rate of heat transmission through walls, floors and roofs. In some cases the areas or rates of heat loss through windows are also governed but there are instances where, although the properties of areas of opaque construction are controlled, unlimited window areas are permitted. These legislative standards are clearly oriented towards health and comfort rather than energy conservation.

Many regulations make reference to condensation and include provisions governing cold bridges, vapour barriers and minimum ventilation rates.

5.13 Thermal capacity of construction

The thermal capacity of construction is taken into account in some countries, particularly in Europe. Lightweight, low thermal capacity walls are required to have higher values of thermal insulation than heavier, high thermal capacity construction. This measure reduces the rate at which lightweight buildings will cool during night-time temperature setbacks and must be regarded as a measure related to health and comfort rather than energy conservation. In some cases of lightweight construction either thermal storage heating is required or central heating must be provided with input at night.

5.2 Energy conservation

Legislation specifically for energy conservation presents particularly difficult problems. There are two separate aspects. One is the operation of the building, the other is the construction of the building itself or modifications of the fabric such as increased insulation.

5.21 Reduction of air temperature

As has been demonstrated above, reduction of air temperature has a very significant effect on energy conservation. In principle it would be possible to require that winter temperatures should be reduced. Legislation to this effect has been introduced in both the UK and France. In the UK there is no attempt at enforcement and, except in buildings where energy saving for economy is taking place, there is little observable result. In France checks are made by inspectors. It would be interesting to know what degree of saving has been achieved and whether the cost of the inspectorate is justified by energy saving. In reducing winter temperatures air conditioned buildings must be carefully considered. In many cases the design results in the operation of the cooling plant during most of the heating season. In cases of this sort reducing the air temperature would merely increase the waste of heat.

5.22 Other factors

Other factors related to use of the building could be as important as reduced temperature in achieving energy savings. They are control of ventilation and efficient operation of the plant. Apart from the Swedish Building Norm due to come into effect in 1977, which has recommendations for checking plant efficiency, there appears to be no legal control of either of these important aspects of waste.

5.23 Timescale for legislation effects

Legislation for buildings is rarely retrospective except in some aspects of safety from fire. New requirements for energy conservation are limited to new buildings. This means that the effect of this type of provision is not likely to have any significant effect on energy conservation in the short term while, in the long term, it is very difficult to know now what conditions will be fifty or sixty years in the future, which is the order of delay before a major portion of the building stock could be influenced. Diagram 5.1 enables one to predict the period of time over which building legislation is likely to be effective. The diagram shows the percentage of the building stock which has been renewed over given periods of time for renewal rates of 1 and 2 per cent. In recent times new housing in the UK has proceeded at a rate of 2 per cent of the stock each year but this is likely to be reduced in the future. Pre-war studies in some cities indicate a renewal rate close to 1 per cent which seems to be a reasonable assumption for future predictions. The straight lines show the situation which would result if only old buildings were renewed. In practice new buildings are almost as likely to be replaced or demolished as old ones and the curved lines show the rate at which the housing stock would be renewed over periods of time assuming that buildings are replaced without any regard for their age. A realistic value is clearly some-

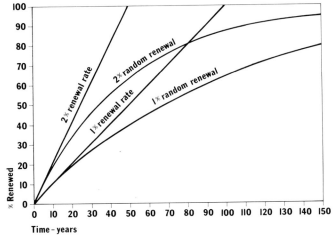

5.1 *The percentage of building stock which has been renewed over given periods of time for renewal rates of 1 and 2 per cent.*

where between the straight line and the curve, in all probability very much nearer to the curve than the straight line.

In the present context the diagram may be used to estimate how soon legislation on the basic form and construction of buildings would be effective in terms of energy conservation. If it is assumed that it is necessary to have $\frac{2}{3}$ of the building stock conforming with the legislation for the influence of the legislation to be regarded as effective it will be observed from the diagram that on a 1 per cent random renewal rate this situation will not be achieved for nearly 100 years. On this basis it is apparent that, if it is to be effective, building legislation must anticipate the requirements of nearly 100 years time in the future.

5.24 Proper targets for legislation

It is hardly surprising, therefore, that although many countries have revised their standards of insulation which, all other things being equal, must give rise to some saving of energy, there is virtually no evidence of any precise energy saving target. Such targets must involve, to be relevant, not only the heat losses from buildings, but also the number and size of buildings giving rise to the losses. There can be no doubt that more efficient use of the building stock would have a greater impact on energy conservation than insulating an excessively large and inefficient utilised stock of buildings.

This would, however, result in some reduction of standards of space and convenience and, at present, few people other than a small band of dedicated enthusiasts would be prepared to accept any real constraint, however small, in order to save energy.

The main factors which would be involved in this type of legislation may be summarised:
1. Number of buildings and efficiency of utilisation of space.
2. Grouping of buildings into blocks to minimise losses.
3. Multiple use of buildings.
4. Energy sources.
5. Heat losses through the fabric and through ventilation.
6. Internal air temperatures.
7. Efficiency of operation of plant.
8. Control and response of plant considered in conjunction with the thermal behaviour of the building and the pattern of occupancy and use.
9. Design of the building and fenestration to take advantage of solar gain in winter.
10. Avoidance of summer overheating.

Note on air conditioning

In many countries air conditioning will be used very much less in the future except in the case of large volumes or specialised environments. Careful design of the fabric to limit peak temperatures can be effective over a wide range of climates and building uses and legislation to ensure that this is done may well be made in future.

In climates where air conditioning is required generally the points above are equally relevant.

5.3 Comparison between UK, US and Sweden

In the light of these considerations it is relevant to consider in more detail examples of legislation governing the thermal performance of buildings and, in so far as it is possible, the reasons for the particular developments. The United Kingdom, the United States and Sweden offer an interesting comparison.

5.31 The UK

In the UK, prior to 1963, building byelaws were made by local authorities guided by model examples issued by the central

government offices. After 1963 National Building Regulations prepared by the central government were adopted. One set of regulations covered England and Wales, with the exception of London, and a similar set of regulations covered Scotland. Prior to 1953 some individual local authorities imposed their own requirements for thermal insulation with the object of combatting condensation and improving comfort, but generally buildings did not have to conform with any thermal standards. Between 1953 and 1972 there were very modest requirements for insulation of roofs, walls and floors. These requirements increased somewhat in stringency during the period but could normally be met by conventional construction without any major provision for insulation. One significant change occurred as a result of the introduction of the new building regulations in 1963. It was the omission of the requirement that every habitable room should be provided with either a flue or a vent to the fresh air. The intention of this change was to improve comfort conditions but, in conjunction with changes in the types of heating installation, methods of construction and patterns of occupancy the result was a marked increase in the incidence of condensation. In 1974 new regulations for insulation of dwellings were introduced requiring a very much higher standard of insulation and, by means of specifying an overall average U-value for walls including windows, the area of glazing was controlled. Although they were introduced during the current energy crisis there were no powers, at the time, to make building regulations for energy conservation and the regulations of 1974 are based on the needs of health, and in particular thermal comfort and freedom from condensation. These regulations are still current.

They are shown in table VIII.

Table VIII Maximum U-value for different elements of housing buildings (from table to regulation F3)
Figures in brackets are the old U-values where values had previously been given.

Element of building	Max U-value W/m² °C
External wall	1·0 (1·70)
Perimeter walls, including windows	1·8
Wall between a dwelling and a ventilated space (ie with permanent vents exceeding 30 per cent of wall boundary area)	1·0
Wall between a dwelling and a partially ventilated space (ie with permanent vents not exceeding 30 per cent of wall boundary area)	1·7
Wall between a dwelling and any part of an adjoining building to which Part F is not applicable	1·7
Wall or partition between a room and a roof space including that space and the roof over that space	1·0
External wall adjacent to a roof space over a dwelling, including that space and any ceiling below that space	1·0
Floor between a dwelling and the external air	1·0 (1·42)
Floor between a dwelling and a ventilated space	1·0
Roof including any ceiling to the roof or any roof space and any ceiling below that space	0·6 (1·42)

One of the most significant aspects of these new regulations is the requirement that the average U-value of perimeter walls including windows should not exceed 1·8 W/m²°C. The effect of this regulation on the planning and economics of house design is considerable. Diagram 5.2 shows that for a detached house the maximum area of single glazing must not exceed 17 per cent which is considerably less than is found in many dwellings of this sort. The use of double glazing makes reasonably free use of windows possible but at considerable penalty in cost. Diagrams 5.3 and 5.4 show that semi-detached and end of terrace houses are affected similarly but less acutely than detached ones. The design of terrace houses will be little affected by the regulation.

5.32 The US

In the United States of America each state government has building codes oriented to public health and safety. There are

24

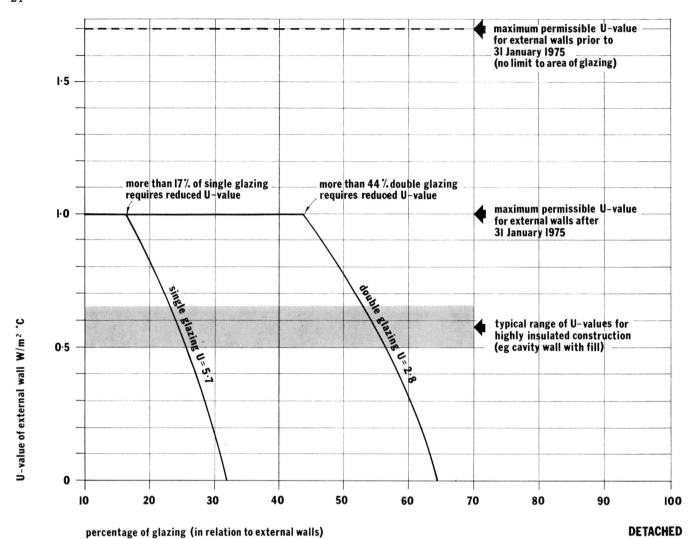

5.2 *Relationship between wall U-value and percentage of glazing possible in perimeter walls of detached houses using the maximum U-value* | *for perimeter walls of 1·8 W/m² °C and of 1·0 W/m² °C for external walls.* | *Example: a U-value of 1·0 W/m² °C will allow a maximum of 17 per cent of single glazing or of 44 per cent of double glazing.* | *A U-value of 0·5 W/m² °C gives percentage figures of 25 and 56 respectively.*

DETACHED

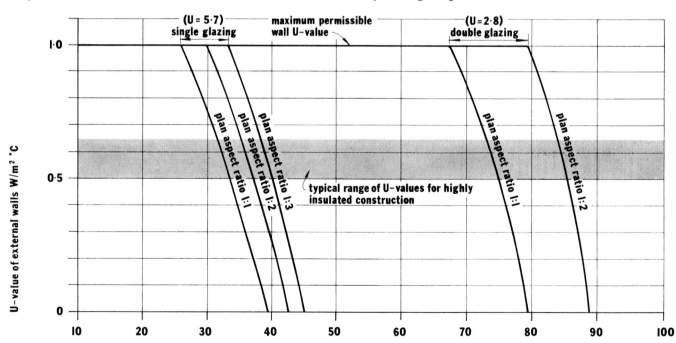

5.3 *Relationship between wall U-value and percentage of glazing possible in external perimeter walls of* | *semi-detached houses and of ends of terraces. (See also 5.5). Example: for a plan form ratio of 1:2 (see 5.5) a* | *U-value of 1·0 W/m² °C will allow a maximum of 30 per cent single glazing or of 79 per cent double* | *glazing. A U-value of 0·5 W/m² °C gives percentage figures of 36 and 85 respectively.*

SEMI-DETACHED & END OF TERRACE

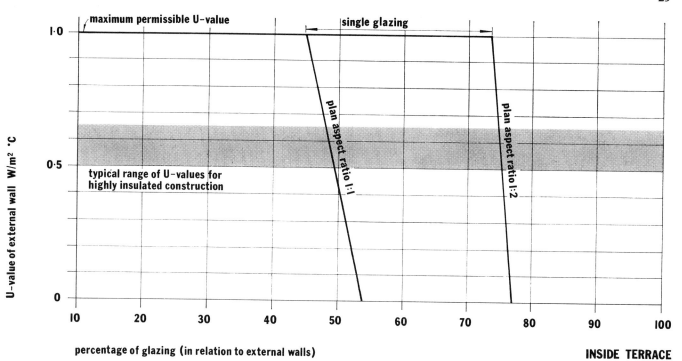

percentage of glazing (in relation to external walls)

INSIDE TERRACE

several National Model Building Codes on which individual building codes are often based. Although some states have adopted provisions related to energy conservation most building codes do not cover this point. The Federal Housing Administration in 1965 included in its minimum property standards requirements to restrict heat losses. They controlled the rate of heat loss for the building in terms of a maximum loss per unit floor area. In relation to energy conservation this type of provision seems eminently sensible. It allows freedom for the designer to work within an overall heat loss. The functions of the Federal Housing Administration were taken over by the United States Department of Housing and Urban Development and in 1973 new minimum property standards for thermal insulation were issued. They specify maximum rates of heat loss through walls, ceilings and floors and are related to different climatic zones by means of varying values for different ranges of degree days. The Department of Housing and Urban Development estimated that the result of the new requirements would be a saving of from 25 to 40 per cent over the previous standards. It is by no means clear why the change in the method of specifying the degree of insulation should have changed. Although some balancing between the values for walls, floors and ceilings is permitted in the new standards it appears to be a less flexible method than the previous one. The tables given in the minimum property standards are too extensive to quote but a comparison between these standards and UK and Swedish ones is given in table IX. Considerable emphasis is given in the minimum property standards to vapour barriers and the prevention of condensation and also to the problems of minimising energy requirements for air conditioning in summer.

5.33 Sweden

In Sweden Svenska Byggnorm 67 gave detailed requirements for thermal performance of buildings which included, not only insulation but also window area and took account of the thermal capacity of walls. In order to cover the range of climatic variations the country is divided into four climatic zones. In zone IV, the warmest southernmost areas, the insulation standards required for brick walls were less stringent than those of the UK in 1974 and the USA in 1973. Higher insulation values were required for lightweight construction where the lower thermal capacity could allow rapid cooling at night with reduced comfort and increased condensation risks. These standards are to be superseded by Supplement No 1 to Svenska

5.4 Relationship between wall U-value and percentage of glazing in terrace houses. (For explanation of aspect ratios, see 5.5). Compared with 5.2 and 5.3 the percentage of glazing allowed may be sufficient to require only single glazing, but for all *plan form ratios shown, the complete external walls could be fully glazed by using double glazing. For a plan form ratio of 1:3 the external walls may be fully single-glazed.*

5.5 Plan shapes, to be read in conjunction with 5.3 and 5.4.

Table IX Comparison of UK, USA and Swedish requirements for the insulation of walls and windows in houses
(Overall thermal conductance inside air to external air, W/m²°C)

Country	Walls	Windows	Notes
UK*	1·0	5·7 (single glazing) 1·8 (double glazing)	Average value for walls including windows, doors etc. not to exceed 1·8
USA‡ one and two family dwellings	0·57		No overall requirement. Higher heat loss rate in some elements may be compensated by lower losses in others
		3·7	
Multi-family dwellings	0·68		
Sweden†	0·3	1·0	No overall requirement but window area is controlled. Higher heat losses are permitted if internal temperatures will not exceed 18°C (0·47 for walls)

* Building Regulations 1974
‡ US Dept of Housing and Urban Development, Minimum Property Standards, 1973 (2500–4450 Degree Day values used for comparison: similar range to UK)
† Svenska Byggnorm 1975 Supplement No 1 (Climate Zones III and IV used for comparison)

Byggnorm 1975 which is to come into effect in 1977. This development increases the standards of insulation, provides for control of ventilation, and lays down standards for heating installations including their management to ensure continuing efficiency. In addition measures are envisaged to control the type of fuel used if this should prove necessary and new buildings are to be required to have space which could be used in case of need for the storage of indigenous fuel.

Table IX compares the insulation values required by SBN 75, Supplement No 1 with the UK and USA values for walls and windows. Account is no longer taken of thermal capacity but window areas are limited to a proportion of the floor area.

6 OTHER IMPORTANT CONSIDERATIONS

6.1 Summer overheating

6.11 Factors which influence peak summer temperatures in buildings

Some of the measures which will improve winter economy can result in summer overheating and care must be exercised to ensure that this does not result. The following notes summarise the factors involved and the histogram illustrates the effects of various control measures upon high and low thermal capacity buildings.

Schedule of factors

The schedule alongside lists, in the left hand column, those aspects of the building design which may be involved in determining the peak temperatures which are reached in buildings in summer; and indicates, in the right hand column, how these may be manipulated to avoid excessive peak temperatures.

Histograms showing relative effectiveness

The histograms on page 29 show, for a selected typical office, the effect of varying several of the design aspects listed in the preceding schedule. The values indicated apply *only* to the specific example described below, and cannot be used directly for general design. They do, however, give a good indication of the factors which are most effective; and in the absence of precise techniques which can be used during the early design stages to aid the development of sensible building concepts, these histograms should be a valuable rough guide. They will be helpful not only in the development of new designs, but also in the modification of existing buildings to reduce summer overheating problems.

The room on which the graphs are based is a south-facing office 4 m × 4 m × 2·85 m high, with 8 m² of single glazed window which is open in the daytime and gives a ventilation rate of 3 air changes an hour. The external walls are of cavity construction with brick outer leaf, and plastered insulating block inner leaf. The internal partitions are lightweight concrete block with plastered finish. Floors are 200 mm concrete. Two occupants are present in the room; and no electric lights are used during the day. There are no obstructions to sunlight.

Two basic cases are considered in the two histograms, both for mid-June conditions.

6.2 considers the office as described above, with solid floor and plaster ceiling: therefore, an enclosure of *high thermal capacity* as described in item 2 of the foregoing schedule. Buildings of high thermal capacity are described as 'slow-response' because their internal temperatures will rise or fall only slowly relative to changes outside — the large storage capacity of the enclosure serves to 'iron out' the fluctuations.

6.3 considers the same office, but with carpet on floor, ceiling tiles below concrete roof, and lightweight partitions. These low-capacity linings transform the originally high-capacity building to one that performs thermally in a way similar to *low-capacity* buildings, as described under item 3 in the

Table X Speed of response of different wall materials		
Material	Thickness required for U-value of 1 W/m² degC	Temperature rise resulting from application of 1 kW for 1 minute
Concrete	830 mm	0·04°C
Brickwork	700 mm	0·06°C
Timber	120 mm	0·68°C
Lightweight concrete	250 mm	0·24°C
Wood wool	83 mm	1·40°C
Fibreboard	42 mm	4·80°C
Expanded polystyrene	25 mm	96·00°C

Schedule of factors which influence peak temperature

Aspect of building	Action to reduce peak temperatures
1 Absorption of heat by external surfaces	Use low-emmissivity surface (white or, preferably, shining)
2 Thermal capacity of walls, floors and roofs and internal partitions	High thermal capacity absorbs heat during the day, instead of transmitting it direct to the building interior; and re-radiates it slowly during the night. Peak temperatures are therefore *reduced*. (Designing for optimum winter conditions may produce conflicting requirement favouring *low* thermal capacity; designer must decide which case takes precedence.)
3 Absorption of heat by internal surfaces of building	Low thermal capacity linings such as carpets or ceiling tiles, in effect convert a building of high thermal capacity to one of low thermal capacity, by preventing heat being absorbed into walls, floors and ceilings. Such linings therefore *increase* peak temperatures, and their summer overheating problems. (But again designing for optimum winter conditions may produce a conflict, by favouring low-capacity linings.)
4 Insulation near outer surfaces of buildings	Less heat from solar radiation will penetrate surface and be transmitted to interior, and more heat lost back to the outside air, therefore peak internal temperatures will be *reduced*.
5 Insulation at or near internal surfaces of buildings	Similar to 3; should be avoided from summer comfort point of view (but may be appropriate for winter conditions).
6 Ventilation of cavities	Can prevent solar heat from reaching interior but normally thought appropriate only for tropical countries.
7 Ventilation of building interior	Higher rates of ventilation *reduce* peak temperatures. Ventilation at night is useful (but may cause security problems).
8 Air movement	Increased air movement will give perceived improvement in comfort.
9 Fenestration: 9a Window area	Reduced window areas have marked effect in *reducing* peak temperatures.
9b Orientation	East and west facing windows are most difficult to control, south facing ones easier, north facing ones no problem.
9c Window glass	Heat absorbing glasses and (to a greater degree) heat rejecting glasses limit solar heat gain, and *reduce* peak temps.
9d Blinds (internal)	Have a useful effect in *reducing* temp.
9e Blinds (external)	Have a marked effect in *reducing* temp.
10 Shading	Traditionally buildings were shaded by trees and other buildings, but with modern buildings such shading is much reduced. Projections, specially provided to give shade on windows, can be very useful; particularly for south facing windows where horizontal projections are highly effective. (It is curious that while many architects propose that deciduous trees should be used for acoustic screening—for which they are ineffective—few use them for solar shading. Properly sited and sized, they can be extremely effective.)
11 Internal heat gains: 11a People	Not significant in summer conditions.
11b Lights	Use minimum lighting levels, and where appropriate extract air through light fittings to remove heat.
11c Processes and other gains	Insulate heat sources; and extract air at source to remove heat.

schedule. As table X shows, linings of materials such as expanded polystyrene will transform a building into one of

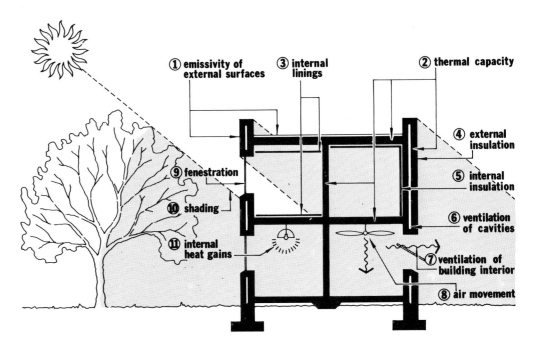

6.1 *Key diagram to the factors which influence peak temperatures attained in buildings, and which the designer may therefore manipulate to avoid summer overheating. The numbered points are more fully discussed on a schedule page 27*

① emissivity of external surfaces
③ internal linings
② thermal capacity
④ external insulation
⑤ internal insulation
⑨ fenestration
⑩ shading
⑥ ventilation of cavities
⑪ internal heat gains
⑦ ventilation of building interior
⑧ air movement

very quick thermal response (carpets on floors will have a generally similar effect).

In each of the two basic cases a range of individual variations has been examined, and the resulting peak temperatures in mid-June compared by means of the histograms. It can easily be seen which are the most effective. The effects of combinations of methods have not been investigated; and it must be repeated that these results apply only to the example described above—they cannot be used for general quantitative design, except for an exactly similar case. But they provide an indication of the general lines along which design should proceed, without the need for time-consuming calculations, and should therefore be helpful to designers.

The histograms show clearly that peak summertime temperatures can be greatly reduced by various window screening devices which limit solar intrusion.

Alternative methods

Window screening should not be seen as the only, or even the most obvious way of dealing with summer overheating. Other methods, as diagrams **6.2** and **6.3** show, can also be effective and should be considered. But there are disadvantages to each; and these are summarised below.

● *Increased ventilation rate* needs to be of a very high order to achieve results comparable with window screening—perhaps 20 changes an hour (instead of the more normal 2 or 3) in the example analysed here. This will be difficult to achieve in many conditions, and may create unacceptable problems of dust and noise intrusion.

● *Reduced window area* is certainly an option to be seriously considered in designing *new* buildings; although the effect of smaller windows upon daylighting, and upon useful solar heat gain in winter (particularly on south-facing facades) may be significant contra-indications. In the case of *existing* buildings, it will seldom be feasible to reduce window size without harming appearance.

● *Shading the window* by building projections is another important option in *new* buildings, which will be more difficult to apply in the case of *existing* ones. As the histogram shows, it can be very effective in the case of south-facing facades; but where windows face east or west, and the sun shines in at a variety of angles (including very low ones), this method will be ineffective.

● *High thermal capacity* (which implies omitting carpets to

concrete floors; ceiling tiles of low thermal capacity beneath concrete slabs; and thermal insulation applied to the inner surfaces of walls and roofs) can also be effective. But there is likely to be a conflict with the requirements of designing for winter conditions (where low thermal capacity may well be preferable); and there can be no certainty that future occupants or building owners will not neutralise the anticipated effect by installing carpets, ceiling tiles, or inner-surface insulation.

● Few of these objections apply to *window screening*. It is effective; can be applied to new as well as existing buildings; will leave building owners or occupants considerable freedom in preferred window size, glazing type, and application of internal linings to the building, without overheating problems; and need not create conflicts with winter performance. Also, provided the right type of device is chosen, low-angle sun (on east or west facing facades) can be dealt with as well as high-angle (on south-facing walls).

6.2 The menace of condensation

Insulation applied to an inner face presents obvious problems in obtaining a durable finish but, in view of the potential savings (which would be greater in the case of rooms with more than one exposed wall), it seems important to be able to solve the problem. There is, however, a potentially much greater difficulty which appears to be little understood but which, if not taken into account, could produce a major disaster not dissimilar to others which have occurred in the building field in recent history, owing to concentration upon one aspect of design without thought for the other consequences. This is the problem of condensation. Most insulation materials have a very open texture. This contributes to their effectiveness as thermal insulators, but many such materials have a very low vapour resistance. The effect of applying an insulating material of this sort to the inner face of a wall is to reduce very substantially the temperatures outside the insulation on what was the surface of the wall. The passage of vapour is, however, little affected and it is quite possible for a wall which gave no trouble prior to insulation to suffer from condensation after it has been insulated. Even insulation in the cavity can have this effect.

Attempts may be made to control condensation by means of

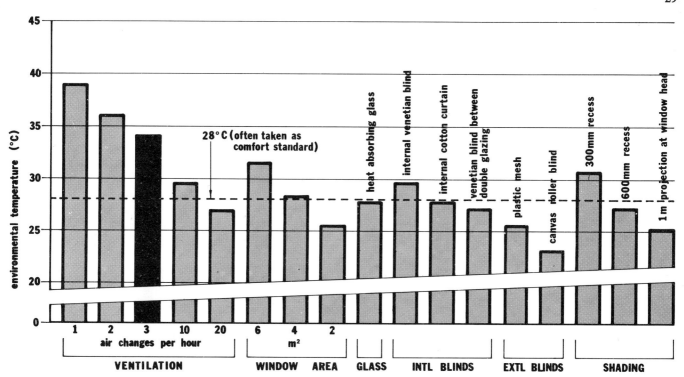

6.2 *Histogram comparing relative effectiveness of various measures which may be taken to reduce peak summer temperature, for specific room described on page 2/7. Black bar shows that room as originally designed (with south-facing clear glass, unscreened window of 8 m²; and ventilation rate of 3 ac/h) will attain 34°C. Each of the grey bars shows what temperature would be if one specific aspect of design were to be modified (eg increasing ventilation to 20 ac/h reduces temperature to 27°C; or holding ventilation at 3 ac/h but reducing window size from 8 m² to 2 m² reduces temperature to 25·5°C).*

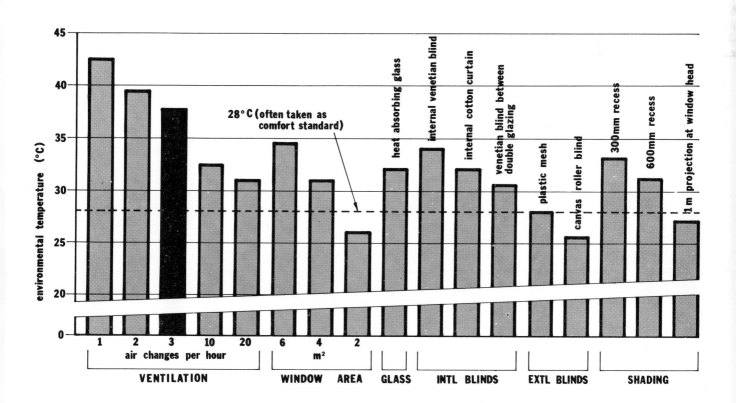

6.3 *Histogram for same room as analysed in **6.2** but assuming that building has been effectively changed from slow-response to fast-response by addition of carpets on floor, and low-capacity ceiling tiles to slab soffit. Note how internal temperature at 3 ac/h rises from 34°C for heavyweight construction, **6.2** to almost 38°C for 'lightweight', **6.3** (ie heavyweight modified by linings). Grey bars again indicate how the 38°C for 3 ac/h may be modified by taking a range of measures.*

vapour barriers and, while this is in principle possible, experience appears to demonstrate that, with current methods and standards of workmanship, failures cannot be ruled out. Very careful thought must be given to condensation in any proposals for insulation in buildings. There is at present very little design advice available on the problem, but current work at BRE is expected shortly to result in authoritative guidance being published.

The graphical method of steady state analysis described in three articles in *The Architects' Journal* (see AJ 19.5.71 p 1149–59 and 26.5.71 p 1201–08) can be used to demonstrate the condensation problems which may arise as a result of reduced ventilation rates or the application of an inappropriate type of insulation inside a wall. Many types of insulation which have excellent thermal resistance have very low vapour resistances.

The first case, **6.4**, shows conditions which might prevail in a house with a simple brick cavity wall. No condensation is demonstrated. The second case, **6.5**, shows how the increase in internal moisture content due to a reduction in the ventilation rate can give rise to condensation. The third case, **6.6**, shows the conditions which may arise when a layer of internal thermal insulation, having negligible vapour resistance is applied. The effects of both insulation and reduced ventilation is demonstrated in **6.7**

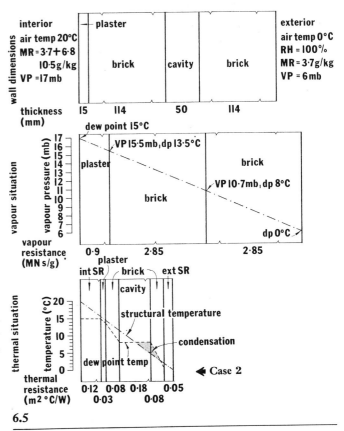

6.5

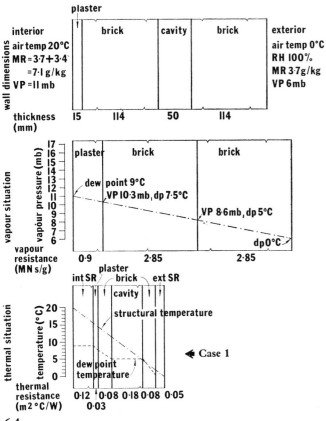

6.4

Key: MR = mixing ratio (ie weight of water present in a sample containing unit weight of dry air in grammes per kilogramme, g/kg)
VP = vapour pressure (here expressed in millibars, mb, but metric SI unit is normally N/m²)
RH = relative humidity (the mass of water vapour present in a sample of air expressed as a percentage of the mass of water vapour which would be present in the sample if it were saturated at the same temperature)
SR = surface resistance of wall (in m² degC/W)
VR = vapour resistance (in meganewton seconds per gramme, MNs/g)

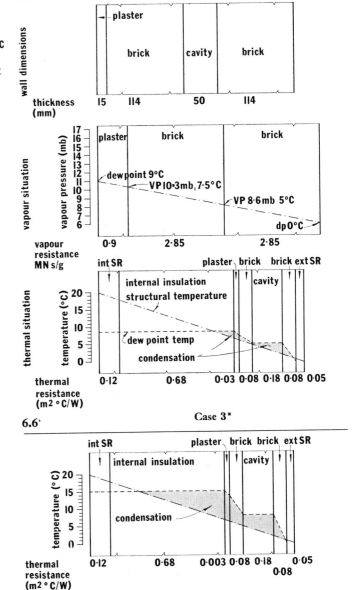

6.6

6.7

* Insulation on inner face of wall has inadvertently been omitted from top diagram.

7 THERMAL DESIGN

7.1 Conclusions from foregoing chapters

The studies described have been concerned with the nature and scale of possible energy savings in the heating of buildings. The appropriateness of any particular method, the actual energy saving, the capital (and energy) investment required and the decision, must depend upon the circumstances of particular cases. It is clearly desirable, however, to attempt to draw some conclusions about the relative values of various methods of energy conservation which are the responsibility of the designer.

In a number of instances direct generalised comparison can be made. In others, such as solar gain, direct comparison could only be made in the context of a specific problem. The case for attention to shape and solar gains is, however, entirely clear. In all new buildings careful thought should be given to shape, orientation and fenestration so that, within the limits of acceptable design, the best arrangement is made to minimise heat losses by careful design of building shape and window sizes and by orientation and siting to maximise gains during the heating season (summer overheating must be checked and avoided). Except in housing where these considerations are an important factor, the potential energy gains are not a large percentage, but since no significant cost or other penalty is involved the return upon investment is infinite and all architects should automatically take this into account in design. It is important to remember that the energy consumption in heating buildings is so vast that even a fractional saving represents an important national resource.

7.11 Time switches and thermostats

Shape and insulation are considerations for new buildings. In both new and existing buildings adjustments can be made to the thermal response of the building (mainly insulation and its position in walls and roofs) and to the control of the heating installation. Tables VI and VII (p16) assume that the installation is effectively thermostatically controlled. Lack of such control gives considerable waste. In domestic buildings appropriate controls are not expensive relative to the increase in comfort and economy, and should be used in all cases. The tables do, however, demonstrate dramatically the very great potential savings which can be achieved by time switching the heating installation to operate only when really required, and to the less great but also significant effect of improving speed of thermal response of the fabric (in this example improved response is achieved by varying the position of the insulation. Inside insulation gives fastest response). Increasing insulation may be worthwhile since even small savings are important, but it is not highly effective in the situations studied except in the case of continuous heating and constructions of very low basic insulation. Apart from hospitals and similar buildings there seems every reason to outlaw continuous heating. When looking at tables V and VI, it is important to remember that, if the new proposals for U-values are adopted, all new walls will have approximately the U-value for the wall studied with 25 mm of insulation. Fortunately measures for energy saving are not usually mutually exclusive but it does appear that the most effective building measure is the time switch, followed by thermostatic control and fast response for the building itself and the heating installation.

7.12 Cost, community and clothes

There are important factors in energy conservation which are quite outside the control of the architect but which are nevertheless very significant to considerations of building design. Probably the most significant of these is the cost constraint on heating which is very real for many people. Studies at BRE have shown that a large percentage of domestic users govern their heat consumption not by the thermostat but by the meter. If additional insulation is used in these cases, the result is likely to be improved comfort rather than energy saving. This may be a meritorious result in some ways but does not save energy.

Multiple use of buildings is clearly an important way of saving not only energy but also capital cost. It is something the community should consider.

Increased clothing is a very effective means of reducing temperature required in buildings with considerable potential energy saving. This may be thought a retrograde step to be avoided but there is no doubt that the use of energy in buildings is a balance between many factors and clothing is clearly one of these. In recent years cheap energy has resulted in lighter clothing. It would be illusory to imagine that now energy is more expensive clothes will not, over a period, become adjusted. In principle very great economies of energy could be made in buildings by the wearing of heavier clothes.

7.13 Rank order of energy savings

Tables XI and XII show comparisons of some energy saving measures with equivalent insulation and design choices of equal effectiveness in relation to their notional costs.

Table XI Effects of various energy saving techniques
Ways of saving energy are given in the first column, and the equivalent thickness of insulation needed to achieve the same result, in the second column.

Strategy for saving energy	Insulation to walls and roof required in addition to double glazing throughout
Reduce ventilation rate from two air changes/hour to the minimum acceptable value of one air change/hour as in *IHVE Guide*	Infinite insulation would still be insufficient
Improve clothing of least clothed occupant from 0·32 clo to 1·25 clo (that is jackets and trousers for women as well as men)	127 mm of foam polyurethane slab
Reduce volume/person from 15 m³ to the legal minimum requirement (or in existing buildings, increase the density of occupation)	82·5 mm of foam polyurethane slab

Table XII Comparison of design alternatives showing effect of choice on building cost

Equally energy effective design choices	Probable effect on building construction costs
Double glaze throughout and insulate walls and roof with 39 mm of foam polyurethane slab	Considerable increase
Reduce volume/person from 15 m³ to 11·3 m³ (legal minimum for factories)	Substantial reduction
Increase clo value from 0·32 to 0·66 (ie put on an extra woolly)	No change
Make full use of intermittent occupation (9 am to 5 pm). This would involve thermally 'light' construction, slightly larger capacity for heating installation and time switched thermostats	No significant change

The possibility that increased insulation, if it is effective, may be used to improve comfort standards, rather than to save energy, demonstrates a principle that needs to be borne in mind when one considers energy saving. The first law of thermodynamics states that while energy may be changed from one form to another, it can be neither created nor destroyed. In energy conservation terms this law can be restated to say that *while energy may be used for a variety of different purposes, it can be neither created nor saved.* If energy is saved in a dwelling, the resources which would have been devoted to its purchase are likely to be redeployed and it is difficult to discover ways of redeployment which do not at some stage involve use of energy. A householder who saves money on energy in his home will use the unspent money for other purposes, all of which ultimately will result in energy consumption, for instance in transport, manufacture, or fertilisers for food. Forced saving as a result of taxation will result in redeployment of the same resource by the government rather than the individual, but whether this is devoted to security, public works, education, health or diplomacy, ultimately energy will be used. If the use of energy by the householder is limited as a result of high prices charged by the producers, then the same resource will be used by the producers, also ultimately resulting in energy consumption.

The only way of saving energy would appear to be for the householder, government or producer to make a conscious decision at the time of saving it to give up the unused resource. If this is not done, the resource will be utilised for other purposes, all of which finally result in the consumption of energy.

It seems that only two approaches are effective. One is consciously to forego whatever resource is saved by conserving energy; in other words it must be saved but not invested or made available for other use. The other method is to weight decisions about which source of energy to use by rationing or taxation so that indigenous rather than foreign sources are chosen. This has little energy-saving value but may be very significant economically.

Every effort should be made in building to design efficiently, since poor and wasteful results cannot be accepted by conscientious building design professions, despite the evidence to the contrary in the recent past. Clearly there can be no saving of energy if it is directly wasted in buildings. On the other hand, overall energy saving is a wider and more complex problem which cannot be solved within the building context alone.

7.14 Action is possible now

It is often said that energy cannot be saved in existing buildings and that the time lag before new buildings would have a significant effect is too long to warrant urgent action. Throughout most of history buildings served the purpose for which they were designed for long enough to justify the design. Now functions change so fast that the original intention for the building has often changed before it leaves the drawing board. A whole new form of transport has developed and serves buildings that were built before the motor car was conceived. But if we now attempt to invest vast resources in buildings to suit the motor car our efforts are likely to be crowned by its disappearance and replacement by different forms of transport. Certainly if present predictions are even vaguely correct, oil as a fuel for heating buildings will have disappeared quite early in the life of buildings being erected now. The architectural profession has a responsibility to meet the intellectual and design challenges presented by the problems of time scale. In relation to energy conservation, however, this problem of time scale does not present the difficulties that might be supposed. In existing buildings, the following items, discussed mainly in Chapter 4, can be completely implemented: multiple use (4.12), increase of population and reduction of volume per person (4.13), increase of clothing and reduction of internal temperature (4.21), and thermostatic control (7.11). Insulation (4.17), ventilation (4.18), thermal response (4.19, consequential upon 4.17), and solar control devices of curtains and shutters (4.20) can be partially implemented.

Energy saving in building does, therefore, offer possibilities of early economy in existing as well as in new buildings.

Although thermal design should be undertaken as a whole we are concerned at present with the importance of the building form and fabric. Decisions about form and fabric can, at one extreme, almost eliminate the need for a heating installation and at the other render necessary not only large-scale heat input but also cooling and air conditioning plant in the summer.

Before design can take place there must be some comparative evaluation of the factors involved in energy economy. There has been no national attention given to this problem since the era of cheap energy. A very small number of institutions have maintained interest and developed skill in this field.

Some of the aspects have been adequately explained and need no further clarification. Those where additional explanation is needed are amplified on the following pages.

7.15 Questions of design

Five questions

The thermal behaviour of buildings is a function of their basic form, fabric, fenestration and orientation. It is something that must be considered from the earliest stages of design; it is not possible to design without reference to thermal considerations and subsequently apply a thermal 'cosmetic' treatment. Present design practice was under considerable criticism before the recent involuntary reassessment of the importance of energy. The importance of design is clear and it is necessary in practice to consider five questions:

1 What design decisions involving energy must be made?
2 What is the importance of each decision?
3 Who should make these decisions?
4 At what stage of design must the decisions be made?
5 What data is available to guide thermal decisions?

7.16 Design decisions

Aspects of building directly affecting thermal performance and their energy significance are summaried on pages 32-3.

Caution

Reduction of ventilation rate can bring about condensation as can the provision of internal insulation of high thermal but low vapour resistance (typical of many insulation materials). This problem must be checked during design.

Heavyweight or lightweight construction

The requirements for avoiding overheating during the summer from solar radiation are different, in many cases, from the requirements for economy of winter heating. There is an urgent need for research into the optimum balance between winter and summer requirements. As yet no balance has been established. Designers must use their judgement. At present the sensible approach to the problem appears to be to design for winter economy while checking that excessive summer temperatures will not occur.

Use	*Encourage multiple use*	In appropriate circumstances (particularly educational buildings) multiple use will give better economy of operation and may save whole buildings.
	Reduce comfort temperature	Inevitably associated with wearing more clothes. One additional pullover can have a significant effect. Present clothing levels are made possible by cheap energy. Increased energy costs will lead to a new balance of clothing and heating.
Siting	*Avoid exposed sites*	Increased energy requirements for buildings on an exposed site are mainly due to higher wind speeds giving reduced U-values and increased ventilation rates. Elevated sites can be colder and more subject to rain and mist.
	Avoid noisy and polluted sites	Where excessive noise or atmospheric pollution is present windows cannot be opened and mechanised ventilation (and probably air conditioning) will be required.
Planning	*Reduce volume*	For a given occupancy the smaller the volume of the building which can be achieved the better the energy economy. Loss through the fabric is reduced when the area of external skin is reduced and the ventilation loss is diminished also. The heat gain from lighting, occupants and processes will, in a smaller building, contribute a much greater proportion of the heat required. Apart from operational savings there is a substantial saving in the energy required for basic construction.
	Use economical shape	Cubical shapes with flat facades have smaller surface areas and consequently smaller heat losses than elongated or elaborately configured shapes.
	Group individual buildings together	Small individual buildings will have greater heat losses than the same buildings combined into a single block. Semi-detached houses have lower heat losses than detached, and terraces are better than semi-detached.
	Avoid open plan	Open plan dwellings mean that the whole volume must be heated whereas with separate rooms differential control of temperature is possible and ventilation losses may be reduced.
	Locate heat sources centrally	Boilers and flues and any other heat source should be located centrally so that heat is contributed to the other rooms in the dwelling and not lost to the exterior. (Often with unfortunate consequences, as in flue condensation.)
Construction	*Provide adequate insulation*	Structural insulation not only reduces heat losses but also raises the internal surface temperatures which will improve comfort and may even permit the use of lower air temperatures.
	Provide appropriate thermal response	If completely continuous heating is to be provided no special consideration of the response of the structure is needed. Except for hospitals and prisons, however, few buildings have continuous heating. High thermal capacity walls and floors lose heat during the period when the heating is off and require substantial pre-heating periods to come back to comfort conditions when the heating is turned on. Low thermal capacity linings are economical of energy in most cases and critically important when heating is very intermittent.
Fenestration	*Reduce window sizes*	Even double glazing has three times the rate of heat loss of an external wall. To conserve energy, window areas should be kept to a minimum. Curtains or shutters for use at night cut down heat losses at the most critical times.
	Consider window orientation	Solar heat gain through windows is a significant factor which varies considerably with orientation. Unobstructed, south facing windows, curtained at night gain as much solar heat as they lose during the winter. East and west windows gain some heat, north windows very little.
Energy source	*Select energy conserving fuel*	1 shows typical efficiency of distribution and utilisation for various types of fuel. The availability of fuel and the suitability of particular types for the building should be considered together with the related primary energy requirements.
	Utilise waste or environmental heat	In appropriate circumstances, not usually domestic, various heat recovery measures can be taken to take waste heat from exhaust air or flue gases and make it available for heating. Heat pumps may extract heat from the surrounding environment and thus provide economical energy.

| Installations | *Choose distribution system and emitters* | The system for distributing heat, if central heating is used, and the emitters in rooms should be chosen to give a response fast enough to give economy in use. Hot water radiator systems respond less quickly to changing demand than thermostatically controlled fan convectors. The consequent balance of economy and comfort must be weighed in design. |
| | *Provide effective control* | Heating systems should be provided with three controls: time controls to enable the installation to be turned off or down at night, thermostatic controls to govern heat output and zone controls to isolate areas not in continuous use. |

7.17 Importance of each decision

In practice, the feasibility and cost of particular thermal measures may be very different for different designs. So it is impossible to lay down any rules, or even rank ordering, for thermal factors which will be universally applicable. Designers must be familiar with the basic principles and exercise appropriate judgement.

Usually, and particularly in existing buildings, there are limited funds for energy conservation and it is not always sensible to tackle the major causes of heat loss but rather to select measures giving the best saving with available resources. In many domestic buildings the major heat loss is through the walls. Walls are, however, difficult and expensive to insulate while roof insulation is usually easy and cheap to install.

Leaving aside consideration of cost and feasibility in specific situations, it is possible to set out a rank order, table XIII, of the effectiveness of various thermal measures for dwellings. The logic of the table may be demonstrated by comparing the effects of changing internal temperature, reducing volume and improving insulation. Improving insulation is of second order effectiveness because although it will reduce fabric losses with the same internal temperature ventilation losses will remain the same.

It is possible to demonstrate, in quantitative terms, the typical degree of importance of most of the design decisions scheduled in question 1.

7.18 Who should make these decisions?

At present decisions governing the thermal performance of buildings are made by architects, though not usually explicitly taking into account the thermal consequences. Generally the decisions about form, materials and fenestration which inexorably govern the thermal environment are taken on grounds of planning, construction, appearance, and economy. In the era of cheap energy this balance may often have been considered appropriate. It is important to remember, however, that the low priority given to thermal performance was not a considered response to the availability of cheap energy but a basic gap in technological expertise and professional coverage.

Decisions by default

In the mid-nineteenth century, when the present pattern of building design professions was established, table XIV, a very limited range of building materials was available. Walls were massive, window openings small, buildings were low by present standards and well protected by trees and other buildings. This gave very specific thermal properties to the vast majority of buildings. Similarly, there was little need of sophisticated decisions about the types of heating installation.

Table XIV Founding of professional institutions

Institution	Founded
RIBA	1834
IOB	1834
RICS	1868
IHVE	1897

If it could be afforded, a gravity, low pressure, hot water space heating system was used. If not, open fires were the only practical alternative.

Architects did not have to consider the thermal properties of the building they designed since no changes could be made. Engineers needed only to consider thermal properties related to the prediction, on a steady state basis, of the maximum rate of heat loss. The answer required, provided it was not too small, did not have to be at all precise. The present pattern of professional responsibility and education was established during this period. The engineering analysis of thermal properties of buildings has grown more complex but it is still orientated towards design of plant rather than the design of the building. Architects rarely analyse thermal behaviour. It has been an era in which no-one made adequate design decisions about the thermal performance of buildings.

The future

It is not easy to postulate who should make the decisions which will be inevitable in the future. The simplest solution would be to return to a very slowly changing technology where economic factors and trial and error development would lead to universally recognised solutions and again no decisions would be required. The rapid changes in technology and in living patterns prohibit this traditional pattern. Unlike the past when very limited choice of materials and heating

Table XIII Rank order of design factors affecting thermal economy in winter

Type of analysis required to enable design decision to be made	Typical rank order of the significance for energy saving of the factors influencing heat loss in dwellings governed by decisions taken during the design. The order is indicative of the potential effect of each factor. The actual importance and the feasibility of each case must be considered during design.		
	1st order	*2nd order*	*3rd order*
Steady state	Grouping	Insulation	Shape
	Comfort temperature	Ventilation	Planning
	Volume		
Dynamic	Zoning and time control of heating installation	Speed of thermal response of building fabric	Speed of response of heating installation
		Fenestration and orientation	

systems together with more stereotyped patterns of living gave little opportunity for choice and even less for incorrect choice, the present variation of building thermal properties, performing of installations and requirement for heating provide the opportunity for major mistakes. Thermal performance must be assessed and decisions made.

Although it is a service that the community will clearly demand of its building design professions it is not at all clear which of the professions will undertake this task. Neither architects nor engineers are, at present, trained to design buildings for optimum thermal performance. It is clear however, that unlike many of the past situations where the need for a particular type of analysis has led to the establishment of a new, separate, professional group, the design of thermal properties is a basic and integral part of the design of the building. It involves decisions about planning, form, materials, fenestrations and orientation that cannot be separated from basic design.

7.19 Timing of decisions

The thermal performance of the building itself is governed by the basic elements of floors, walls, windows, roofs and by the form, fenestration and planning. Thermal design is, therefore, an integral aspect of basic design and must be considered when basic design decisions are taken. Failure to take thermal considerations into account does not mean that the decisions are postponed, only that they have been taken blindly and at random. 7.1 shows the main thermal factors and the typical design stages when thermal decisions are taken.

It will be seen that some important aspects are probably determined by the client before the architect is consulted. The consideration of several more is mainly during the feasibility stage. It is at the sketch design stage, however, that most thermal matters are settled. Very few are of such a nature that significant variations can be made after the sketch design, unless the designer is prepared to make the overall reconsideration of the design which is likely to be required.

7.20 Availability of data

A point of fundamental importance is also demonstrated in 7.1. Most of the decisions affecting thermal performance are inevitably taken at a stage in design before it is possible to make any close evaluation of their consequences. The designer's responsibility is, therefore, very great. His earliest thoughts must be informed by understanding of the thermal consequences.

The only way to overcome this problem even partially is to develop several different early design thoughts to a degree which permits of evaluation. Their consequences can then be compared. Even here, however, it is only possible to evaluate a limited number of possibilities which have been selected by the designer. The fundamental need and responsibility for designers of buildings to have a very thorough appreciation of thermal behaviour is apparent.

7.21 Energy savings in existing buildings

It is often suggested that, since only 2 per cent of the housing stock is renewed each year, no effort to conserve fuel in the design of buildings can be effective. This is far from the case. The large amount of energy which is devoted to buildings means that even over a short period of time savings of considerable magnitude can be made in new buildings.

Existing buildings can also make major energy savings. All the aspects marked with asterisks in 7.1 can be modified in existing buildings, and the improvement of control systems which is easy in existing buildings is one of the most effective energy saving measures.

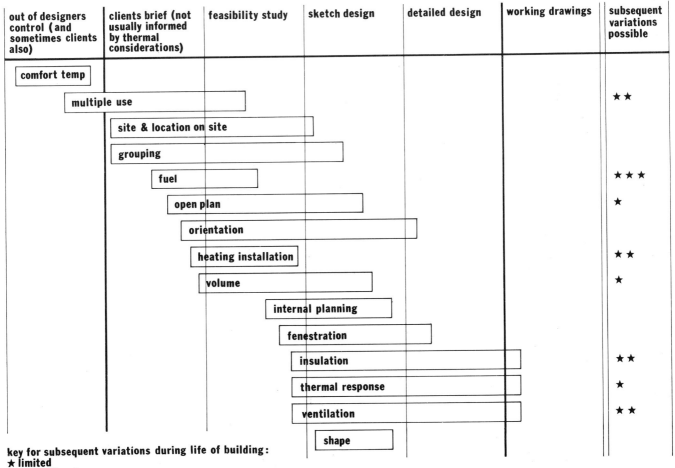

key for subsequent variations during life of building:
★ limited
★★ significant
★★★ complete change possible

7.1 Sequence of thermal design decisions. *In some cases internal planning can be varied.*

8 CALCULATION METHODS

8.01 Effect of England and Wales building regulations

There used to be little need for thermal analysis early in design. Before the building regulations second amendment which came into operation on 31 January 1975* the main requirement was a wall U-value of 1·7 W/m²degC or less. Walls could be fully glazed.

New standards for housing now apply to the shell: walls, floor and roof; and to elements between spaces which are not necessarily heated simultaneously: intermediate floors, partitions and party walls. The limiting of U-values to 1·0 W/m²degC for wall material or 1·8 W/m²degC with glazing requires careful consideration of glazing area (or the real economics of double glazing) and necessitates a balance between thermal, lighting and other aspects of fenestration.

The building regulations are nominally to protect health, for example by combating condensation. New powers now exist, however, to legislate specifically for energy conservation. These will probably lead to standards for other building types and perhaps more stringent standards for housing. Architects will have to be more precise (and knowledgeable) in predicting thermal performance from the outset of design, to check particular constructions, configurations of form, orientation, glazing area and so on.

Thermal control in summer
Current legislation covers heating but not summer overheating. However, several government departments are recommending maximum summer temperatures in briefs for department buildings. Estimating peak summer temperatures involves study of window area, orientation, shading of windows by other buildings, thermal capacities of materials and types of finish. Decisions may be taken early in design.

Seasonal energy requirements
Though not yet subject to control, for example by energy budgets, seasonal energy requirements will inevitably become more important to building owners and occupiers as fuel costs rise. Changing cost differentials between fuel costs and costs of fabric and plant will change priorities in the brief.

Plant
In many cases space must be allocated for plant early on, since it takes up a significant volume of the building. Estimation of approximate plant size is therefore an essential part of an outline design.

The purpose of calculation
With the shift in emphasis on thermally important decisions from final detail to early stages comes a need for quick numerical checks on thermal performance without the need for frequent consultations. The methodology and data in this chapter provide step by step procedures for these checks which may become necessary for every building. The procedures provide for all relevant factors so that the architect can follow a procedure without needing to familiarise himself once more with all the details.

Current regulations
Current U-value standards, as given in the building regulations amendment, are shown in table XV.

* See chapter 5.

Table XV Maximum U-values from building regulations second amendment

Element of building	Maximum U-value of any part of element W/m²degC
External wall	1·0
Wall between a dwelling and a ventilated space	1·0
Wall between a dwelling and a partially ventilated space	1·7
Wall between a dwelling and any part of an adjoining building to which Part F is not applicable	1·7
Wall or partition between a room and a roof space, including that space and the roof over that space	1·0
External wall adjacent to a roof space over a dwelling, including that space and any ceiling below that space	1·0
Floor between a dwelling and the external air	1·0
Floor between a dwelling and a ventilated space	1·0
Roof, including any ceiling to the roof or any roof space and any ceiling below that space	0·6

8.02 U-value calculation

Most existing tables of U-values were prepared before the new building regulations amendment and usually give ranges of values too great to be currently practicable. So U-values must often be calculated from basic data.

A standardised basis for U-value calculation is set out in BRS Digest 108 *Standardised U-values* (August 1969). It establishes standard assumptions and gives necessary data. The following procedure takes these assumptions into account.

Worksheets
This U-value calculation and other procedures include worksheets compatible with them. These are suggested as standard documents for the architect's office. The values entered in some of the worksheets are from the worked examples which accompany procedures.

The worksheets are as follows:
8.1 Basic U-value procedure.
8.2 Insulation for given U-value calculation.
8.3 Maximum rate of heat loss calculation.
8.4 Plant sizing calculation.
8.5 Seasonal heat requirement calculation.
8.6 Peak summer temperature.
8.7 Area of glazing.
8.8 Inner surface temperature.
8.9 Condensation.

8.1 Basic U-value procedure

1 Using the worksheet, **1**, list the materials including cavities which form each layer of the wall, floor or roof in column 1. Give the thickness in metres in column 2.

2 Using table XVI choose the appropriate exposure and enter the value of external surface resistance in the first row of column 4.

3 Using table XVII select cavity resistance values and enter in column 4 on the appropriate row.

4 Using table XVIII or XX select conductivity 'k' values for each material and enter in column 3.

5 Divide the thickness (column 2) by the conductivity (column 3) and enter the result (the resistance R) in column 4.

6 From table XIX select the appropriate internal surface resistance and enter it in the last row of column 4.

7 Sum column 4 and enter total.

8 Calculate reciprocal of 7, ie 1/R. This is U-value in W/m² degC.

JOB TITLE		COMPONENT EXTERNAL WALL	
WORKSHEET NO		U-VALUE CALCULATION	
ELEMENT OF CONSTRUCTION	THICKNESS 't' metres	CONDUCTIVITY 'k' W/m deg C	RESISTANCE R = t/k m² deg C/W
EXTERNAL SURFACE RESISTANCE*			0.053
1 BRICK EXTERNAL SKIN	0.105	0.84	0.125
2 CAVITY (VENTILATED)	0.050	–	0.18
3 LIGHTWEIGHT CONCRETE BLOCK	0.10	0.25	0.40
4 LIGHTWEIGHT PLASTER	0.015	0.17	0.088
5			
6			
INTERNAL SURFACE RESISTANCE			0.12
		TOTAL RESISTANCE	0.966
*in case of partition or floor this will be 'internal surface resistance'.		U-VALUE (1/R)	1.03

8.11 Example

Consider a wall of normal exposure and construction: external skin brickwork, 50 mm cavity; 100 mm lightweight concrete block; 15 mm lightweight plaster finish.

The completed worksheet is shown above. Where materials can vary considerably in properties, eg when described as lightweight, reference is better made to table XX than the typical values suggested by table XVIII.

1 List elements of construction: brick, cavity (ventilated), block and plaster.

2 External surface resistance for high emissivity* in normal exposure from table XVI. 0.053.

3 Ventilated cavity resistance from table XVII, 0.18.

4 Brick and block conductivity values from table XVIII are

* See footnote to tables XVI and XVII.

Table XVI External surface resistance for various exposure conditions

Building element	Surface emissivity*	Surface resistance (m²degC/W) for various exposures		
		Sheltered	Normal	Severe
Wall	High	0.080	0.053	0.027
	Low	0.106	0.062	0.027
Roof	High	0.070	0.044	0.018
	Low	0.088	0.053	0.018

* Emissivity is high for all normal building materials including glass. It is low for unpainted or untreated metallic surfaces such as aluminium or galvanised steel.

Tables XVI to XXI and XXIII are from P. Burberry, *Environment and services*, Batsford.

Table XVII Standard thermal resistance of airspaces (including internal boundary surface)

Nature of airspace	Surrounding construction	Thermal resistance m²degC/W
Ventilated (minimum 20 mm thick)	Airspace between asbestos-cement or black metal cladding with unsealed joints, and high emissivity* lining	0.16
	As above, with low emissivity surface facing airspace	0.30
	Loft space between flat ceiling and unsealed asbestos-cement or black metal cladding pitched roof	0.14
	As above with aluminium cladding instead of black metal, or low emissivity upper surface on ceiling	0.25
	Loft space between flat ceiling and unsealed pitched, tiled roof	0.11
	Loft space between flat ceiling and pitched roof lined with felt or building paper, with beam filling	0.18
	Airspace between tiles and roofing felt or building paper	0.12
	Airspace behind tiles on tile-hung wall	0.12
	Airspace in cavity wall construction	0.18

		(Heat flow horizontal or upwards)	(Heat flow downwards, eg floors)
Unventilated	50 mm thick, high emissivity*	0.11	0.11
	50 mm thick, low emissivity	0.18	0.18
	20 mm thick, high emissivity	0.18	0.21
	20 mm thick, low emissivity	0.35	1.06
	High emissivity corrugated and plane sheets in contact	0.09	0.11
	Low emissivity multiple foil insulation	0.62	1.76

* Emissivity is high for all normal building materials including glass. It is low for unpainted or untreated metallic surfaces.

Table XVIII Typical thermal properties of various materials

Material	Density kg/m²	Thermal conductivity (k) W/m°C	Thermal resistivity (1/k) m²degC/W
Asbestos-cement	1600	0.36	2.78
Asphalt	1700	0.50	2.00
Brickwork, common	1700	0.84	1.19
Compressed straw slabs	260	0.09	11.1
Concrete:			
ballast	2000–2400	1.0–2.0	0.5–1.0
cellular	450–950	0.10–0.28	3.6–10
clinker	1500–1700	0.33–0.40	2.5–3.0
foamed slag	650–1100	0.14–0.25	4.0–7.2
vermiculite	350–950	0.07–0.28	3.6–14.3
Cork board	140–320	0.05–0.06	16.7–20
Expanded polystyrene	15–30	0.03–0.04	25–33
Fibreboard	460	0.05	20.00
Glass	2500	1.02	0.98
Glass wool	50	0.035	28.6
Hair felt	80	0.04	25.00
Hardboard	640	0.10	10.00
Plasterboard	960	0.16	6.25
Plaster, dense	1300	0.50	2.00
Plywood	530	0.14	7.14
Roofing felt	1000	0.20	5.00
Slates	2600	1.80	0.57
Stone	2000–2800	1.30–2.80	0.36–0.77
Tiles, roof	1900–2250	0.83–0.94	1.06–1.20
Timber	640	0.14	7.14
Vermiculite	80–100	0.03–0.04	25–33
Wood wool slabs	470–800	0.08–0.14	7.14–12.5

Table XIX Internal surface resistance

Element	Resistance m² deg C/W
Walls	0.12
Floors (downward heat flow)	0.15
Ceilings (upward heat flow)	0.11

0.84 and 0.25 respectively. The value for lightweight plaster, 0.17, is taken from table XX.

5 Divide 't' by 'k' for brick, block and plaster.

6 Internal surface resistance from table XIX, 0.12.

7 Total resistance, 0.966.

8 U-value is 1/0.966 = 1.03.

Table XX Thermal conductivity of masonry materials for different moisture contents (covers brick, lightweight concrete and dense concrete)

Density dry kg/m³	Thermal conductivity: W/m² deg C for following moisture contents (percentage by volume)							
	1	2	3	5	10	15	20	25
200	0·09	0·10	0·11	0·12	0·15	0·16	0·18	0·19
400	0·12	0·13	0·15	0·16	0·19	0·22	0·24	0·25
600	0·15	0·17	0·18	0·20	0·24	0·27	0·29	0·32
800	0·19	0·21	0·23	0·26	0·31	0·34	0·37	0·40
1000	0·24	0·27	0·30	0·32	0·39	0·43	0·47	0·51
1200	0·31	0·35	0·38	0·42	0·50	0·56	0·61	0·66
1400	0·42	0·47	0·52	0·57	0·68	0·76	0·82	0·89
1600	0·54	0·60	0·66	0·73	0·87	0·98	1·06	1·14
1800	0·71	0·79	0·87	0·96	1·15	1·28	1·39	1·50
2000	0·92	1·03	1·13	1·24	1·49	1·66	1·80	1·95
2200	1·18	1·32	1·45	1·59	1·91	2·13	2·31	2·50

Note: 1 per cent—brickwork protected from rain; 2 per cent—concrete protected from rain; 5 per cent—brick or concrete exposed to rain. Protected implies inner skin of cavities, backing to external cladding and external walls. Exposed implies external skin of cavity walls and solid external walls. (Based on BRS Digest 108, August 1969.)

Table XXI U-values for many types of construction

Basic construction	Thickness mm	External finish	Internal finish	Resistance m²degC/W	U-value W/m²degC over 1·7	1·7–1·0	below 1·0
Solid brickwork	105			0·3	3·3		
	,,		15 mm hard plaster	0·33	3·0		
	,,		15 mm lightweight plaster	0·39	2·5		
	,,		25 mm cavity and foil backed 10 mm plasterboard	0·67		1·5	
	,,		25 mm exp polystyrene + 10 mm plasterboard	1·12			0·9
	,,	Tile hanging	15 mm lightweight plaster	0·55	1·8		
	,,	Tile hanging	25 mm cavity + 10 mm plasterboard	0·7		1·4	
	220			0·43	2·3		
	,,		15 mm hard plaster	0·46	2·1		
	,,		15 mm lightweight plaster	0·52	1·9		
	,,		25 mm cavity + 10 mm plasterboard	0·67		1·5	
	,,		25 mm cavity + foil backed 10 mm plasterboard	0·77		1·3	
	,,		25 mm exp polystyrene + 10 mm plasterboard	1·25			0·8
	,,	Tile hanging	15 mm lightweight plaster	0·68		1·5	
	,,	Tile hanging	25 mm cavity + 10 mm plasterboard	0·83		1·2	
	,,	Tile hanging	25 mm cavity + 10 mm foil backed plasterboard	0·95		1·1	
	,,	Tile hanging	25 mm exp polystyrene + 10 mm plasterboard	1·41			0·7
	335			0·55	1·8		
	,,		15 mm hard plaster	0·58		1·7	
	,,		15 mm lightweight plaster	0·64		1·6	
	,,		25 mm cavity + 10 mm plasterboard	0·79		1·3	
	,,		25 mm cavity + foil backed 10 mm plasterboard	0·91		1·1	
	,,		25 mm exp polystyrene + 10 mm plasterboard	1·37			0·7
	,,	Tile hanging	25 mm cavity + 10 mm plasterboard	0·95		1·0	
	,,	Tile hanging	25 mm cavity + foil backed 10 mm plasterboard	1·07			0·9
	,,	Tile hanging	25 mm exp polystyrene + 10 mm plasterboard	1·53			0·7
Cavity brickwork 105 mm brick inner and outer leaves, 50 mm cavity	260		15 mm hard plaster	0·64		1·6	(1·0)*
	,,		15 mm lightweight plaster	0·7		1·4	(0·9)
	,,		25 mm cavity + 10 mm plasterboard	0·85		1·2	(0·8)
	,,		25 mm cavity and foil backed 10 mm plasterboard	0·97		1·0	(0·7)
	,,		25 mm exp polystyrene + 10 mm plasterboard	1·43			0·7 (0·6)
	,,	Tile hanging	15 mm lightweight plaster	0·86		1·2	(0·8)
	,,	Tile hanging	25 mm cavity + 10 mm plasterboard	1·01			1·0 (0·7)
	,,	Tile hanging	25 mm cavity + foil backed 10 mm plasterboard	1·13			0·9 (0·7)
	,,	Tile hanging	25 mm exp polystyrene + 10 mm plasterboard	1·59			0·6 (0·5)
Cavity wall 105 mm brick outer leaf, 50 mm cavity, 100 mm aerated concrete inner leaf	255		15 mm hard plaster	0·96		1·0	(0·7)
	,,		15 mm lightweight plaster	1·05			1·0 (0·7)
	,,		25 mm cavity + 10 mm plasterboard	1·17			0·9 (0·6)
	,,		25 mm cavity + foil backed plasterboard	1·29			0·8 (0·6)
	,,		25 mm exp polystyrene + 10 mm plasterboard	1·75			0·6 (0·5)
Cavity wall 100 mm aerated concrete inner and outer leaves 50 mm cavity	250	Tile hanging	15 mm hard plaster	1·45			0·7 (0·5)
	,,	Tile hanging	15 mm lightweight plaster	1·51			0·7 (0·5)
	,,	Tile hanging	25 mm cavity + 10 mm plasterboard	1·66			0·6 (0·5)
	,,	Tile hanging	25 mm exp polystyrene + 10 mm plasterboard	2·24			0·5 (0·4)
Solid aerated concrete wall	100		15 mm hard plaster	0·66		1·5	
	,,		15 mm lightweight plaster	0·72		1·4	
	,,	15 mm lightweight plaster	15 mm lightweight plaster	0·81		1·2	
	,,	Tile hanging	15 mm hard plaster	0·82		1·2	
	,,	Tile hanging	15 mm lightweight plaster	0·88		1·1	
	,,	Tile hanging	25 mm cavity + 10 mm plasterboard	1·03			1·0
	,,	Tile hanging	25 mm cavity + foil backed 10 mm plasterboard	1·15			0·9
	,,	Tile hanging	25 mm exp polystyrene + 10 mm plasterboard	1·61			0·6

Basic construction	Dimensions	Edge condition	U-value W/m²degC over 1·7	1·7–1·0	below 1·0
Solid floor	30 × 30	Four exposed edges			0·26
	30 × 15				0·36
	30 × 7·5				0·55
	15 × 15				0·45
	15 × 7·5				0·62
	7·5 × 7·5				0·76
	3 × 3			1·47	
	30 × 30	Two exposed edges at right angles			0·15
	30 × 15				0·21
	30 × 7·5				0·32
	15 × 15				0·26
	15 × 7·5				0·36
	7·5 × 7·5				0·45
	3 × 3			1·07	

(continued on next page)

Table XXI (*continued from previous page*)

Basic construction	Dimensions	Construction detail	U-value W/m²degC over 1·7	1·7-1·0	below 1·0
Suspended timber floor	30 × 15	Bare or with lino, plastics or rubber finish			0·39
	30 × 7·5				0·57
	15 × 15				0·45
	15 × 7·5				0·61
	7·5 × 7·5				0·68
	3 × 3			1·05	
	30 × 15	Carpet or cork finish			0·38
	30 × 7·5				0·55
	15 × 15				0·44
	15 × 7·5				0·59
	7·5 × 7·5				0·65
	3 × 3				0·99
	30 × 15	Any surface finish plus 25 mm quilt over joists			0·30
	30 × 7·5				0·39
	15 × 15				0·33
	15 × 7·5				0·40
	7·5 × 7·5				0·43
	3 × 3				0·56

Basic construction	Direction of heat flow	Construction detail	U-value W/m²degC over 1·7	1·7-1·0	below 1·0
Intermediate floor	Up	Timber: 20 mm on joists with 20 mm plasterboard ceiling		1·6	
	Down			1·4	
	Up	Concrete: 150 mm with 50 mm screed	2·7		
	Down		2·2		
	Up	Concrete: 150 mm with 50 mm screed plus 20 mm wood	2·0		
	Down			1·7	
	Up	200 mm hollow tile floor	1·9		
	Down			1·6	
	Up	200 mm hollow tile floor with 20 mm wood		1·5	
	Down			1·3	
Pitched roof		Slates or tiles on battens with felt backing, and plasterboard ceiling	1·9		
		As above but with 50 mm glass fibre insulation between joists			0·54
Flat roof		20 mm asphalt on 150 mm concrete	3·3		
		As above with ventilated cavity + 50 mm glass fibre on 10 mm plasterboard			0·53
		20 mm asphalt or screed on 50 mm wood wool, with ventilated cavity and 10 mm plasterboard ceiling		1·16	
		As above with 50 mm glass fibre on plasterboard			0·44
		Bituminous felt or boarding with ventilated cavity and 10 mm plasterboard ceiling	1·85		
		As above with 50 mm glass fibre on plasterboard			0·51
Windows (normal exposure)		Wood frame: single glazed	4·3		
		double glazed	2·5		
		Metal frame: single glazed	5·6		
		double glazed	3·2		
		(with thermal break in frame)			
Glazing (normal exposure)		Roof glazing	6·6		
		Horizontal light with skylight (ventilated space between)	3·8		

* Note U-values in brackets take into account 25 mm glass fibre slab built into cavity.

8.2 Insulation for given U-value calculation

A calculation of the insulation needed to be added to a proposed (or existing) construction to bring it up to required U-value standard.

8.21 Example

Suppose a wall were proposed which by the previous calculation had a U-value of 1·63 W/m²degC. The thickness of polystyrene insulation required to bring it up to 1·0 W/m²degC can be calculated:

1 'k' value from table XVIII is 0·035.
2 Reciprocal of U required 1/1 = 1.
3 U-value of proposed wall = 1·63.
4 Reciprocal of 3 is 1/1·63 = 0·613.
5 Subtract 4 from 2 is 1·0-0·613 = 0·387.
6 Multiply 5 by 1 is 0·387 × 0·035 = 0·0135 m (13·5 mm).

JOB TITLE	COMPONENT EXT. WALL
WORKSHEET NO	CALCULATION OF THICKNESS OF INSULATION REQUIRED TO ACHIEVE DESIRED U-VALUE

DESIRED U-VALUE 1·0 W/M² DEG C.

1 ESTABLISH CONDUCTIVITY 'k' VALUE OF PROPOSED INSULATION MATERIAL (from information given in table 1V) .... 0·035

2 CALCULATE RECIPROCAL OF U-VALUE THAT IS TO BE ACHIEVED (reciprocal = 1/U) .... 1·00

3 CARRY OUT U-VALUE CALCULATION OUTLINED PREVIOUSLY FOR WALL WITHOUT ADDED INSULATION (use worksheet no 1) .... 1·63

4 CALCULATE RECIPROCAL OF 3 ABOVE (reciprocal = 1/U) .... 0·613

5 SUBTRACT THE ANSWER TO 4 FROM THAT OF 2 (this is the thermal resistance that the wall lacks) .... 0·387

6 MULTIPLY THE RESULT OF 5 BY THE CONDUCTIVITY 'k' OF THE PROPOSED INSULATION MATERIAL, 1. (the answer is the required insulation thickness, in metres) .... 0·0135 M

8.3 Maximum rate of heat loss calculation

Architects will not normally do full engineering calculations but calculation of room input required will indicate emitter sizes. Maximum overall building heat loss can be used in estimating plant size and seasonal heat requirement.

8.31 Procedure

1 Using the worksheet below, list in column 1 the types of construction which make up the envelope of room or building.
2 Enter surface area in appropriate row of column 2.
3 Enter U-values for all surfaces in column 3 (these can either be computed as indicated earlier or taken from table XXI where appropriate).
4 Decide internal and external design temperatures (°C) using table XXII, and enter difference in column 4. When walls separate the space considered from an unheated one, the conventional temperature difference adopted is 5°C.
5 For ventilation, using the bottom of the worksheet, specify volume or number of people. Using table XXIII decide on an appropriate rate of heat loss and enter in column 3. Enter the temperature difference between inside and outside in column 4.
6 For each row (for construction and ventilation) multiply columns 2, 3 and 4 together and enter in column 5.
7 Sum losses for construction as total fabric loss rate.
8 Sum total fabric loss rate and ventilation loss rate to give total heat loss rate.
Note: the value found in 8 will be increased if intermittent heating is to be used (usually by a factor of 1·5 or 2).
For plant sizing it is usual to add a safety factor of 25 per cent and make appropriate allowances if there are heat losses from runs of pipe not inside the building. These losses are generally small but architects in doubt should consult a specialist.

8.32 Example

Consider a house of 130 m² floor area. The approximate overall heat loss is required to size the boiler. Ground floor temperature 21°C and first floor 18°C. External temperature −1°C. Two air changes on both floors. (Bathroom ignored.) This is set out in the worksheet, below left.

Table XXII Internal and external design temperatures

Location	Building type		Temperature °C
Internal	Gallery and museum		20
	Assembly hall and lecture room		18
	Canteen and dining room		20
	Factories:	sedentary	19
		light work	16
		heavy work	13
	Domestic:	living room	21
		bedroom	18
		bathroom	22
		circulation and service spaces	18
	Hotels:	bedrooms	22
		public rooms	21
	Offices		20
	School classroom		18
	Shops		18
External (heating systems designed to have at least 20 per cent output capacity above minimum)	Single storey		−3
	Multi-storey (with solid floors and partitions)		−1

Note: temperatures as proposed by *IHVE guide*.

Table XXIII Recommended minimum fresh air supply to buildings for human habitation

Building type	Rate of fresh air supply	Ventilation heat loss W/°C
Assembly halls	28 m³ per hour per person	9·5
Canteens	28 m³ per hour per person	9·5
Factories and workshops:		
work rooms	22·6 m³ per hour per person	7·7
lavatories and wcs	2 air changes per hour	*
Hospitals:		
operating theatres and X-ray rooms	10 air changes per hour	*
wards	3 air changes per hour	*
Houses and flats:		
bathroom and wcs	2 air changes per hour	*
halls and passage	1 air change per hour	*
kitchens	56 m³ per hour	19
living rooms and bedrooms:		
8·5 m³ per person	20·5 m³ per hour per person	7
11·5 m³ per person	18·5 m³ per hour per person	6·3
14 m³ per person	12 m³ per hour per person	4·1
pantries and larders	2 air changes per hour	*
Places of entertainment	28 m³ per hour per person	9·5
Restaurants	28 m³ per hour per person	9·5
Schools:		
occupied rooms (classrooms, laboratories, practical rooms, etc):		
2·8 m³ per person	42 m³ per hour per person	14·3
5·6 m³ per person	28 m³ per hour per person	8·5
8·5 m³ per person	20·5 m³ per hour per person	7
11·2 m³ per person	18·5 m³ per hour per person	6·3
14 m³ per person	12 m³ per hour per person	4·1
Cloakrooms	3 air changes per hour	*
Corridors, lavatories and wcs	2 air changes per hour	*

* The following data convert air changes to ventilation heat loss per m³

Ventilation rate in air changes per hour	Heat loss per m³ W/m³degC
1	0·34
2	0·68
3	1·02
4	1·36
5	1·70
6	2·04
7	2·38
8	2·72
9	3·06
10	3·39
15	5·09
20	6·79

JOB TITLE				ENVELOPE OR ROOM
WORKSHEET NO				CALCULATION OF MAXIMUM RATE OF HEAT LOSS

FABRIC HEAT LOSS −

1 FABRIC CONSTRUCTION	2 AREA m²	3 U-VALUE W/m² degC	4 TEMPERATURE DIFFERENCE °C	5 RATE OF FABRIC HEAT LOSS (2 x 3 x 4) W
SOLID GRND FLOOR	65	1.0	22	1430
GRND FL. WALLS	75	1.0	22	1650
GRND FL. WINDOWS	18	5.7	22	2227
1ST FLOOR WALLS	79	1.0	19	1501
1ST FLOOR WINDOWS	14	5.7	19	1516
ROOF	65	0.6	19	741

VENTILATION HEAT LOSS −

VENTILATION LOCATION	NUMBER OF PEOPLE; ALTERNATIVELY VOLUME IN m³	VENTILATION RATE PER m³, OR PER PERSON	TEMPERATURE DIFFERENCE °C	RATE OF VENTILATION HEAT LOSS (2 x 3 x 4) W
GROUND FLOOR	150	0.68	22	2244
FIRST FLOOR	150	0.68	19	1836

TOTAL HEAT LOSS RATE (W) ....13,145.......

note − the watt is a measure of heat flow *rate*; for example one watt equals approximately 3600 joules per hour

8.4 Plant sizing calculation

(Approximate for early design stages)

8.41 Boiler room, fuel store, flue

1 Carry out maximum rate of heat loss calculation described previously.
2 For boiler room size and typical layout consult **8.1** and table XXIV.
3 For oil or solid fuel storage size consult table XXV.
4 For flue size consult table XXVI.

8.42 Refrigeration plant

1 Establish approximate cooling loads from the following data:
Perimeter zone (up to 7 m from window):
● 25 per cent glazing—cooling load 120 W/m² of floor area;
● 60 per cent glazing—cooling load 180 W/m² of floor area.
Interior zone (more than 7 m from window):
● cooling load 75 W/m² of floor area.
2 Calculate total cooling load: the sum of floor areas × their cooling loads.
3 Consult table XXVIII for size of refrigeration plant and room and access.

8.43 Air handling plant

1 Consult table XXIII to establish the rate of air change.
2 Multiply rate per person by total population served (or rate per m³ by total volume served).
3 Reduce figures from 2 in m³/h to m³/s (divide by 3600; 1 m³/h = 0·00028 m³/s).
4 Consult table XXVII.

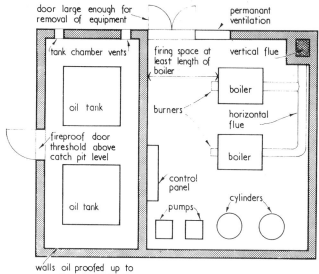

8.1 *Typical layout for medium sized oil fired boiler room. Oil tanks should be as near as possible to boiler plant and at the same level or higher. Note that space required for pipe connections and services access is similar for coal and gas fired boilers.*

Table XXIV Boiler room sizes

Total boiler capacity	Boiler size			Overall size of boiler room m*		
kW	Number of boilers	Size m	Weight when full of water kg	l	w	h
115	2	1·8 × 1·2	770	4·6 ×	3·1 ×	3·1
550	2	2·4 × 1·2	3630	6·1 ×	6·1 ×	3·7
1000	2	3·4 × 1·6	5100	9·1 ×	7·6 ×	3·7
2300	3	3·7 × 1·8	6800	13·7 ×	9·1 ×	4·6
4300	3	2·1 × 4·0	8200	15·2 ×	10·7 ×	5·5

* Note: necessary heights of boiler room.

Table XXV Fuel storage requirements

Total boiler capacity kW	Fuel storage Oil litres	Solid fuel m²	Permanent vent to plant room m²
115	4500	6·5	0·2
550	18 000	26	0·9
1000	32 000	42	1·7
2300	77 000	100	3·7
4300	136 000	186	7·0

Table XXVI Approximate flue dimensions

Total boiler power	Flue size	
kW	Height m	Cross-sectional area m²
115	13	0·1
550	21	0·3
1000	30	0·4
2300	34	0·8
4300	37	1·7

Table XXVII Ventilation plant room sizes

Total air handled m/s³ All air systems	Plant room size length m	width m	height m	Fresh air intake air cooling m²	ventilation only m²	Filter (roll type) width m	height m	depth m
7	8·2	6·6	3·3	1·86	0·47	2·3	1·75	1·2
30	9·8	13·0	4·6	7·9	2·02	7·2	2·3	1·2
Air/water systems (induction systems and so on)								
15	9·8	11·5	3·7	3·7	1·86	2·6 plus 1·3	2·3	1·2
25	9·8	13·0	4·6	6·5	3·25	3·0 × plus 1·7 ×	2·3 3·0 × 3·0 ×	1·2 1·2
60	two plant rooms required, each: 9·8	13·7	4·6	two required, each: 7·5	3·7	two required, each: 6·8 plus 3·6	3·0 3·0	1·2 1·2
110	two required, each: 9·8	26·0	4·6	two required, each: 14·0	7·0	two required, each: 6·8 plus 3·6	3·0 3·0	1·2 1·2

Note: Tables XXIV–XXVII were originally published in the AJ Building Environment Handbook in 1969.

Table XXVIII Sizes of cooling plant and plant room

Cooling load kW	Typical water chillers number	length m	width m	Plant room plan length m	width m	Access for plant width m	height m	Cooling tower length m	width m	height m	Weight of each tower kg	Condenser connections number	Pipe diam mm
120	2 or	1·9	0·8	7·2 or	5·9	0·8 or	2·0	two required, each: 1·5 1·0		1·8	550	2	75
	1	3·2	1·0	7·9	4·3	1·0	2·0	(packaged)					
560	2 or	3·7	1·0	9·0 or	6·6	1·0 or	2·0	two required, each: 1·7 3·0		2·1	1200	2	125
	1	3·8	1·4	10·7	4·6	1·4	2·2	(packaged)					
1050	2	4·1	1·4	10·7	7·6	1·4	2·2	two required, each: 4·5 2·3		2·1	2250	2	200
								(packaged)					
2500	2	4·1	1·4	10·7	9·1	1·7	2·6	two-cell, together: 6·4 7·6		4·0	20 000	2	300
								(site erected)					
5200	2	5·8	3·2	12·8	12·2	2·2	2·0	two-cell, together: 7·0 12·5		4·3	34 000	2	450
								(site erected)					

8.5 Seasonal heat requirement calculation

Estimates of seasonal energy requirement are useful in indicating fuel costs and for comparing designs of heating and/or ventilating systems.

8.51 Degree day method

The degree day method was once used for seasonal heat requirement. From local meteorological records the number of days when the external air temperature fell below a critical level (15°C), and also the average difference between the external temperature and base heating temperature for the whole period, were calculated. The product of these two figures, days × degrees, is the degree day value for the locality. The method is not accurate largely because it ignores heat gains within the building and from the sun.

8.52 HVRA method

The Heating and Ventilating Research Association developed a method based on observation of many buildings which used the number of hours of boiler operation. This evidently gives a realistic estimate for similar buildings, systems and locality. Advice is usually available from consultants for large building types if data are needed for this method.

8.53 BRE method

For housing, architects may bear the responsibility for seasonal energy requirement decisions and a BRE method has existed for many years which is still useful.
Nominal U-values are used except for windows where U-value is reduced to take account of curtaining. A rate of heat loss calculation is based on these values, the length of the heating season (33 weeks), and an average figure of temperature difference between inside and outside depending on the type of heating installation. Data are provided for including heat gains. The BRE method is given here.

Procedure

1 Calculate rate of heat loss using U-values from table XXI and the worksheet, above right, except for:
● U-value for windows 3·4 W/m²degC single glazing;
● roof U-value to be multiplied by 0·75;
● temperature difference to be 10°C for houses with full central heating and 6·7°C for heating mainly confined to the ground floor;
● ventilation loss rate to be taken from table XXIII.
2 Multiply heat loss rate from 1 by 0·02 to convert to GJ.
3 Calculate the heat gains using the worksheet, above right.
4 Subtract gains, 3, from losses, 2, to give the net seasonal energy requirement in GJ.

JOB TITLE		BUILDING
WORKSHEET NO		CALCULATION OF HEAT GAINS (FOR SUBSEQUENT SEASONAL ENERGY REQUIREMENT CALCULATION)
SOURCE	UNIT GAIN GJ	GAINS GJ
WINDOWS	ORIENTATION AREA X GAIN m² GJ	
	SOUTH 0.68	
	EAST & WEST 0.41	
	NORTH 0.25	
PEOPLE	1 GJ PER PERSON	
ELECTRICITY	BASED ON ESTIMATE OF CONSUMPTION. AVERAGE WATTAGE CONSUMED x 0.02 GIVES TOTAL HEAT GAIN IN GJ	
COOKING	GAS 6 GJ ELECTRICITY 4 GJ (UNLESS INCLUDED ABOVE)	
WATER HEATING	2GJ	
OTHER		
TOTAL GAINS (GJ)		

Worksheet for seasonal energy requirement.

5 Multiply the net requirement 4 by a factor representing the efficiency of the heating system:
● open fire   2·5
● central heating   1·4
(includes electric thermal storage)
● on-peak electricity   1·0
This is the gross seasonal energy requirement in GJ.

8.6 Peak summer temperature

Overheating in highly glazed buildings is widespread, especially where external noise restricts window opening and hence natural ventilation. But even with external noise, careful design of fabric, finishes, fenestration and orientation may provide comfortable conditions without air conditioning.† The problem affects offices and schools more than houses. Both the DES and DOE are moving towards specifying peak acceptable temperatures.

8.61 BRE method

BRE has developed the admittance method for calculating peak summer temperatures. The details and background are described in BRS CP 47/68 *Summer time temperatures in buildings* by A. G. Loudon, and CP 61/74 *Thermal response and the admittance procedure* by N. O. Millbank and J. Harrington-Lynn. The method is for office buildings where the office in question has one external wall, and all other faces abut other offices—each side, behind, above and below. It is possible to extend the use of the calculation to rooms not surrounded by others with reduced accuracy but nevertheless usefully. If lightweight construction is used, solar heat gain through external roofs and walls can be taken into account using section 6 of the *IHVE guide*.

The method employs some terms not previously in general use:

Admittance factor
This represents the extent to which heat enters the surface of materials in a 24-hour cycle of temperature variation. Dense materials take up more heat and have higher admittance values than lightweight ones.

Environmental temperatures
A single value representing:

$$\frac{\text{Air temperature}}{3} + \frac{\text{Mean radiant temperature} \times 2}{3}$$

It is more closely related to comfort than a single air temperature and is coming to be used in calculations for radiant heating. The conventional comfort temperature values are usually just values of air temperature. The same values can be used for environmental temperature.

Solar gain factor
The proportion of incident sun's rays transmitted through glazing to the room. It takes account of angle of incidence, losses in glazing, blinds and other sun control devices.

Alternating solar gain factor
A measure of solar gain through glazing taking into account the effects of the thermal capacity of the structure.

8.62 Admittance procedure

The *IHVE guide*'s procedure is complex; BRE proposes eventually to publish a simplified procedure with data for use at an early design stage. The following step by step procedure fills the current gap, enabling peak summer temperatures to be estimated quickly and reliably with a minimum of study of the method itself. There are three stages:
● establish mean conditions (using the worksheet, page 44);
● determine variation from mean (using the worksheet, page 45);
● from these calculate environmental temperature (using the worksheet, page 45).
The handwritten numbers on the worksheets are for the example given later. To use the worksheets read the labels on boxes, look up tables for appropriate values for the building, enter values and do the arithmetic indicated.

Note
The time of peak internal temperature is not always known. Peak solar gain usually predominates in fixing peak times, especially in highly glazed buildings. It may be necessary to check more than one possible time, eg time of peak solar gain and time of peak external temperature, 8.2, 8.3.
In lightweight buildings all factors can be assumed to act simultaneously. If the construction is heavyweight, the solar gain for two hours before the time under consideration should be used.

Example
Consider an office in the southern facade of an office building.
● Room dimensions: 4 m wide, 4 m deep, 2·85 m high.
● Area of single glazed window: 8 m². (This leaves 3·4 m² of solid external wall.)
● Floor construction: 200 mm concrete with carpet plus plasterboard and ceiling battens on the underside.
● Partitions: lightweight concrete block with plaster finish.
● External walls: brick cavity with insulating block inner leaf and plaster finish, having U-value of 1·00 W/m²degC.
● No artificial lighting used during day.
What is the peak temperature in June at 13·00 h for two people doing light office work? (Note that the graph, 8.2, shows that the highest temperatures are likely to occur in July.)
Following through the worksheets, pp 44-5, shows a mean temperature of 28·5°C and hence fundamental changes are needed.
One of the most effective steps would be to reduce window area and/or to control solar gain through the window by special glass, blinds or shutters. If feasible, additional windows giving through ventilation could be effective. Increasing the mass of the structure would be useful as would removing the carpet and the plasterboard from the ceiling.

JOB TITLE

WORKSHEET NO   CALCULATION OF PEAK ENVIRONMENTAL TEMPERATURE - STAGE ONE.

SOLAR

SELECT DAILY MEAN SOLAR GAIN (table XXIX)	X SOLAR GAIN FACTOR FOR GLAZING (table XXX)	X WINDOW AREA m²	= MEAN SOLAR GAIN W	
161	x 0.77	x 8	= 992	(1)

OTHER GAINS

SOURCES	RATE W	X DURATION hours	÷ 24	= DAILY MEAN GAIN W	
LIGHTING (installed wattage)					
OCCUPANTS (No x Rate/Occupant) table XXXI	2 X 140	x 8	÷ 24	= 93	
			TOTAL.............. 93		(2)

TOTAL

MEAN SOLAR GAIN (1)	+ MEAN DAILY GAIN (2)	= TOTAL MEAN GAIN W	
992	+ 93	= 1085	(3)

FABRIC LOSSES

EXTERNAL ELEMENT	AREA m²	X U-VALUE W/m² degC (table XXI)	= RATE OF LOSS W/degC	
WINDOW	8	x 5.7	= 45.6	
WALL	3.4	x 1.0	= 3.4	
		TOTAL.............. 49		(4)

VENTILATION LOSSES

VENTILATION RATE air changes/h table XXXII	X ROOM VOLUME m³	X 0.33	= MEAN VENTILATION HEAT TRANSFER RATE W/degC	
3	x 45.6	X 0.33	= 45	(5)

FOR RATES OF VENTILATION OVER 2/hour COMPLETE :-

1 ÷ answer to 5	= (a)	0.21 ÷ TOTAL AREA OF INTERNAL SURFACES m²	= (b)
1 ÷ 45	= 0.02	0.21 ÷ 77.6	= 0.003
(a) + (b)	= (c)	1 ÷ (c)	= MEAN VENTILATION HEAT TRANSFER RATE W/degC
0.02 + 0.003	= 0.023	1 ÷ 0.023 = 43.5	(6)

MEAN ENVIRONMENTAL TEMPERATURE

FABRIC LOSS (4)	+ VENTILATION LOSS (5) except if more than 2 changes/hour, then use (6)	= TOTAL RATE OF HEAT LOSS W/degC	
49	+ 43.5	= 92.5	(7)

TOTAL GAIN (3) W	÷ TOTAL LOSS (7) W/degC	= SUBTOTAL	+ MEAN EXT TEMP °C (see graph 6.3)	= MEAN INTERNAL ENVIRONMENTAL TEMPERATURE °C	
1085	÷ 92.5	= 11.7	+ 16.8	= 28.5	(8)

Worksheet and sample data for admittance procedure; stage one, mean conditions. Typed figures are standard for all calculations, those handwritten are for the example given.

		PEAK INTENSITY SOLAR RADIAT'N W/m² (table XXIX)	DAILY MEAN – SOLAR INT. W/m² (table XXIX)	EFFECTIVE = PEAK INPUT	AREA OF X GLASS m	= SUBTOTAL	ALTERNATING X SOLAR GAIN FACTOR (table XXXI)	EFFECTIVE = GAIN SWING W	
SOLAR	VARIATION	540	− 161	= 379	× 8	= 3032	× 0.43	= 1304	(9)

CASUAL GAIN VARIATION — PEAK

CASUAL GAINS AT PEAK HOURS		RATE W	
LIGHTING			
OCCUPANTS TABLE VII	2 × 140	280	
	TOTAL	280	(10)

CASUAL GAIN VARIATION — VARIATION

PEAK (10)	– MEAN (2)	= CASUAL GAIN VARIATION W	
280	– 93	= 187	(11)

AIR TEMP VARIATION

AREA OF GLAZING m²	U-VALUE OF X GLAZING (table XXXI)	HEAT = TRANSMITTED W/ degC	(5) OR USE + (6) IF MORE THAN 2/h	FABRIC PLUS = VENTILATION LOSSES	EXTERNAL AIR X TEMP SWING degC (Figure 8.3)	AIR TEMP = INPUT VARIATION W	
8	× 5.7	= 45.6	+ 43.5	= 89.1	× 7	= 623.7	(12)

TOTAL INPUT VARIATION

SOLAR VARIATION (9)	+ CASUAL VARIATION (11)	+ AIR VARIATION (12)	= TOTAL VARIATION W	
1304	+ 187	+ 623.7	= 2114.7	(13)

INTERNAL ENVIRONMENTAL TEMPERATURE SWING — AREA X ADMITTANCE

INTERNAL SURFACE	AREA m²	ADMITTANCE FACTOR FOR x CONSTRUCTION (table XXXIII)	AREA X ADMITTANCE = W/ degC	
WINDOW	8	× 5.6	= 44.8	
EXTERNAL WALLS	3.4	× 2.9	= 9.9	
PARTITIONS	34.2	× 2.6	= 89.0	
FLOOR	16	× 3.1	= 49.6	
CEILING	16	× 5.8	= 92.8	
		TOTAL	286.1	(14)

SWING

AREA X ADMITTANCE (14)	+ VENTILATION GAINS (5) OR (6)	= SUBTOTAL	
286.1	+ 43.5	= 329.6	(15)

TOTAL SWING IN EFFECTIVE INPUT (13)	÷ (AREA X ADMITTANCE) + VENTILATION GAINS (15)	= SWING IN INTERNAL ENVIRONMENTAL TEMPERATURE degC	
2114.7	÷ 329.6	= 6.4	(16)

STAGE THREE — PEAK ENV TEMP

MEAN ENVIRONMENTAL TEMPERATURE (8)	+ SWING IN INTERNAL ENVIRONMENTAL TEMPERATURE (16)	= PEAK ENVIRONMENTAL TEMPERATURE degC
28.5	+ 6.4	= 34.9

Worksheet and sample data for admittance procedure; stage two, variation from mean and stage three, peak environmental temperature.

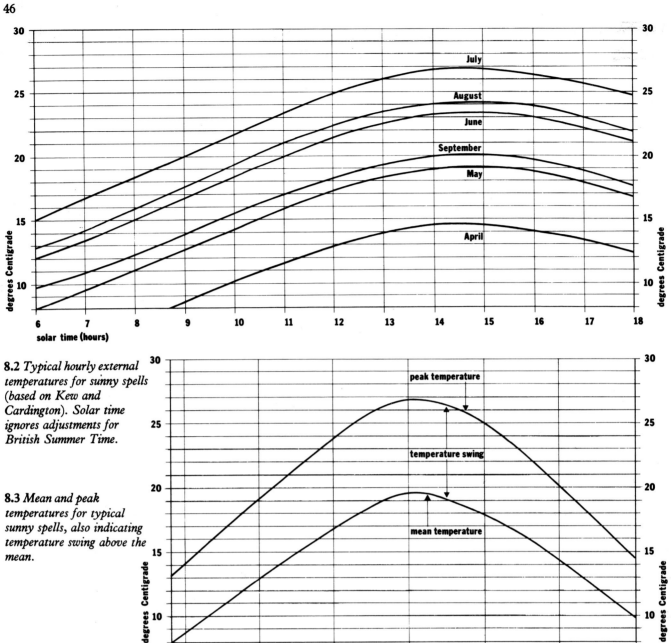

8.2 *Typical hourly external temperatures for sunny spells (based on Kew and Cardington). Solar time ignores adjustments for British Summer Time.*

8.3 *Mean and peak temperatures for typical sunny spells, also indicating temperature swing above the mean.*

Table XXIX Mean and hourly solar intensities on vertical surfaces

Orientation	Month (values are for about 21st day)	24 hour mean	6	7	8	9	10	11	12	13	14	15	16	17	18
East	June	195	595	705	715	660	525	330	130	130	125	110	95	70	55
	May and July	187	550	680	710	660	520	330	130	130	120	105	90	70	50
	April and August	158	400	620	690	650	510	320	120	120	110	105	80	55	30
	March and Sept	112		410	555	555	465	295	95	95	85	70	55	25	
	Feb and Oct	70			330	430	375	250							
South-east	June	190	335	475	595	660	635	540	435	270	125	110	95	75	55
	May and July	197	330	485	630	665	650	580	450	330	120	115	90	70	50
	April and August	191	280	470	610	700	690	620	520	320	110	105	80	55	30
	March and Sept	179		350	575	680	715	670	560	390	195	70	55	30	
	Feb and Oct	146			380	550	650	720	525	380	225	50	30		
South	June	161	55	75	195	340	445	530	585	540	445	340	195	75	55
	May and July	177	50	70	230	395	500	580	610	580	500	395	230	70	50
	April and August	190	30	95	280	405	530	620	650	620	530	405	280	95	30
	March and Sept	206		110	285	450	585	696	715	695	585	450	285	110	
	Feb and Oct	184			230	400	560	670	695	670	560	400	230		
South-west	June	190	55	75	95	110	125	270	435	540	635	660	595	478	335
	May and July	197	50	70	90	115	120	330	450	580	650	665	630	485	330
	April and August	191	30	55	80	105	110	320	520	620	690	700	610	470	280
	March and Sept	179		30	55	70	195	390	560	670	715	680	575	350	
	Feb and Oct	146			30	50	225	380	525	720	650	550	380		
West	June	195	55	70	95	110	125	130	130	330	525	660	715	705	595
	May and July	187	50	70	90	105	120	130	130	330	520	660	710	680	550
	April and August	158	30	55	80	105	110	120	120	320	510	650	690	620	400
	March and Sept	112		25	55	70	85	95	95	295	465	555	555	410	
	Feb and Oct	70								250	375	430	330		

Note: values in 1 m² were calculated using sunpath diagrams and overlays by Peter Petherbridge of BRE.
This table and figures **8.2** and **8.3** appeared in Burberry, P. *Environment and services*, Batsford.

Table XXX Solar gain and alternating solar gain factors for various sun controls

Window construction	Solar gain factors	Alternating solar gain factors	
		Lightweight building	Heavyweight building
Single 4 mm clear glass	0·77	0·55	0·43
Double 6 mm clear glass	0·61	0·47	0·39
Single 4 mm clear glass with internal venetian blind	0·46	0·46	0·43
Double 6 mm clear glass with white venetian blind between glass	0·28	0·25	0·23
Single 4 mm clear glass with external canvas roller blind	0·11	0·09	0·07

Note: the *IHVE guide* 1970, table A8.6, gives the following definitions:
Heavyweight building: solid internal walls and partitions, solid floors and solid ceiling.
Lightweight building: lightweight demountable partitions with suspended ceilings.
Floors either solid with carpet/wood block finish or suspended.
Some of the performance of lightweight construction can be achieved in heavier construction by placing the insulation on the inner face, as illustrated in AJ 4.2.76 p245, figure 2.
The values are from BRS CP 47/68, Loudon, A. G. *Summertime temperatures in buildings*, HMSO, which gives fuller data.

Table XXXI Heat output of young males for various degrees of activity

Activity	Heat output, W
Seated, at rest	115
Light work, office	140
Seated, eating	145
Walking	160
Light bench work	235
Moderate work, or dancing	265
Heavy work	440
Exceptional effort	1500

Note: this is based on table A7.1 from the *IHVE guide*, 1970.
There are no comprehensive data available for children, the old, and others.

Table XXXII Ventilation rates for naturally ventilated buildings on sunny days

Position of opening windows	Usage of windows		Effective mean ventilation rate	
	Day	Night	Air changes per h	Ventilation allowance W/m³ deg C
One side only	Closed	Closed	1	0·3
	Open	Closed	3	1·0
	Open	Open	10	3·3
More than one side	Closed	Closed	2	0·6
	Open	Closed	10	3·3
	Open	Open	30	10·0

Note: this is table A8.4 from the *IHVE guide*, 1970.

Table XXXIII Admittance factors

Construction	Admittance W/m²degC
External walls	
Brick, solid: 105 mm brick 16 mm dense plaster	4·1
220 mm brick 16 mm dense plaster	4·4
105 mm brick 16 mm lightweight plaster	3·1
220 mm brick 16 mm lightweight plaster	3·4
Brick, cavity: 105 mm brick / 50 mm cavity / 105 mm brick / 16 mm dense plaster	4·3
as above but lightweight plaster	3·3
as above but lightweight concrete block inner leaf	2·9
Internal walls	
105 mm brick, 15 mm dense plaster each side	4·5
75 mm lightweight concrete block 15 mm dense plaster both sides	2·6
Two fibre board sheets with cavity between	0·3
Roofs	
150 mm concrete with 19 mm asphalt on 75 mm screed with 15 mm dense plaster ceiling	5·1
50 mm wood wool with 19 mm asphalt on 13 mm screed with cavity and 10 mm plasterboard ceiling	1·5
Windows	
Wood frame: single glazed	4·3
double glazed	2·5
Metal frame: single glazed	5·6
double glazed (with thermal break in frame)	3·2

Floors and ceilings		Floor	Ceiling
Timber	10 mm timber / cavity / 16 mm plasterboard ceiling	0·1	0·3
Concrete	50 mm screed / 150 mm concrete	5·6	5·6
	as above but with wood block or carpet floor finish	3·1	5·8

Note: These values are taken from BS CP 61/74, Millbank, N. O. and Harrington-Lynn, J. *Thermal response and admittance procedure*, HMSO, which gives fuller data.

8.7 Area of glazing

In designing to conform with the building regulations, values may be needed for:
● maximum window area which can be used with a given standard of wall insulation;
● necessary U-value to correspond with a given window area.

8.71 Procedure

Both these values may be obtained by substituting the known values in the basic equation representing the building regulations wall.

$$\frac{xa + yb + 0 \cdot 5c}{a + b + c} = 1 \cdot 8$$

a = area of window.
b = area of external wall.
c = area of party wall.
x = window U-value (W/m²degC: 5·7 for single glazing, 2·8 for double glazing).
y = wall U-value (W/m²degC).

8.72 Example

Calculate maximum permissible area of glazing (a) for given wall U-value.
● Consider a semi-detached house 8 m × 8 m in plan with walls of U-value 0·85 W/m²degC. The walls are 5·2 m high. From these, the area of party wall is 8 × 5·2 = 41·6 m². The area of external windows plus walls is 3 × 8 × 5·2 = 124·8 m²

The basic formula is: $\dfrac{xa + yb + 0 \cdot 5c}{a + b + c} = 1 \cdot 8$

Substituting

$$\frac{5 \cdot 7a + 0 \cdot 85b + 0 \cdot 5 \times 41 \cdot 6}{a + b + 41 \cdot 6} = 1 \cdot 8$$

$5 \cdot 7a + 0 \cdot 85b + 20 \cdot 8 = 1 \cdot 8a + 1 \cdot 8b + 74 \cdot 9$
$3 \cdot 9a - 0 \cdot 95b = 54 \cdot 1$
Since $a + b = 124 \cdot 8$
$ b = (124 \cdot 8 - a)$

Substituting this value for b
$3 \cdot 9a - 0 \cdot 95 (124 \cdot 8 - a) = 54 \cdot 1$
$3 \cdot 9a - 118 \cdot 6 + 0 \cdot 95a = 54 \cdot 1$
$ 4 \cdot 85a = 172 \cdot 7$
$ a = 172 \cdot 7 \div 4 \cdot 85$
$ a = 35 \cdot 6 \text{ m}^2$

This is the maximum window area that can be used.

8.73 Example

Calculate necessary U-value (x) to correspond with a given glazing area.
● Consider a same sized semi-detached house 8 m × 8 m on plan with walls 5·2 m high.
This gives party wall area of 8 × 5·2 = 41·6 m², and area of external windows plus walls of 3 × 8 × 5·2 = 124·8 m².
Suppose that the area of single glazing decided upon is 40 m².
Substituting
$$\frac{5 \cdot 7 \times 40 + y (124 \cdot 8 - 40) + 0 \cdot 5 \times 41 \cdot 6}{40 + (124 \cdot 8 - 40) + 41 \cdot 6} = 1 \cdot 8$$
$\dfrac{228 + 84 \cdot 8y + 20 \cdot 8}{166 \cdot 4} = 1 \cdot 8$
$84 \cdot 8y + 248 \cdot 8 = 1 \cdot 8 \times 166 \cdot 4$
$ 84 \cdot 8y = 299 \cdot 5 - 248 \cdot 8$
$ y = 50 \cdot 2 \div 84 \cdot 8$
$ y = 0 \cdot 59 \text{ W/m}^2 \text{ degC}$

This is the wall U-value required (or lower is acceptable) to comply with the building regulations for 40 m² of single glazing.
Note that the value of y is not always the design value. Using the same figures except for 30 m² of single glazing would give a U-value of 1·14 W/m²degC. Since the walling material U-value must not exceed 1·00 W/m² degC, it (or a lower value) would be used in design.

8.8 Inner surface temperature

Inner surface temperature is important to comfort and is largely controlled by insulation. It can be considerably lower than general room temperature: witness putting a hand near the window of a warmed room on a cold day. This difference between room temperature and inner surface temperature for various constructions is indicated for the dynamic state in the series of diagrams **3.7**, **3.8**.
The chart, **8.4**, relates U-value to inner surface temperature.
For a U-value of 1·8 W/m²degC, the inner surface temperature is about 4°C below the room temperature; this difference is not generally uncomfortable. The chart is more applicable to as yet unregulated buildings such as factories, or parts of buildings such as seating areas in highly glazed foyers.

8.9 Condensation

This is becoming more important as draught stripping and keeping windows closed to retain heat cut down on ventilation. It has been fully discussed in a series of articles in *The Architects' Journal*, by Peter Burberry and Brian Day (CI/SfB (I6), AJ 19.5.71) and methods have been given for determining risk (CI/SfB (I6), AJ 26.5.71).

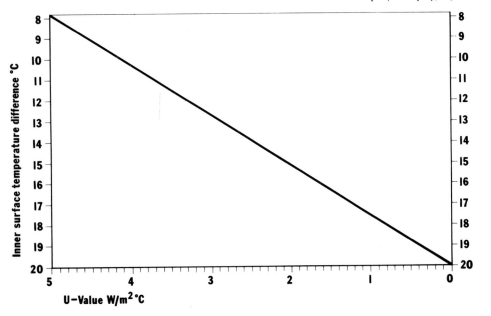

8.4 *Relationship between U-value and inner surface temperature for normal exposure and a difference between external temperature and room temperature of 20°C. The inner surface temperature difference is the difference between external temperature and inner surface. This figure should be subtracted from 20°C to indicate the temperature gradient within the room.*

Inner surface temperature difference °C

U-Value W/m² °C

APPENDIX 1: BLANK WORKSHEETS

Note
Worksheets given in the text are reproduced in this appendix, uncompleted to facilitate office photocopying.

App 1.1 *Worksheet for calculating the U-value of a proposed building element*

JOB TITLE	COMPONENT		
WORKSHEET NO	U-VALUE CALCULATION		
ELEMENT OF CONSTRUCTION	THICKNESS 't' metres	CONDUCTIVITY 'k' W/m deg C	RESISTANCE R = t/k m² deg C/W
EXTERNAL SURFACE RESISTANCE*			
1			
2			
3			
4			
5			
6			
INTERNAL SURFACE RESISTANCE			
		TOTAL RESISTANCE...............	
*in case of partition or floor this will be 'internal surface resistance'.		U-VALUE (1/R)...................	

App 1.2 *Worksheet for calculating the thickness that is required of a proposed insulation material, in order to achieve a desired U-value*

JOB TITLE	COMPONENT
WORKSHEET NO	CALCULATION OF THICKNESS OF INSULATION REQUIRED TO ACHIEVE DESIRED U-VALUE

DESIRED U-VALUE	
1 ESTABLISH CONDUCTIVITY 'k' VALUE OF PROPOSED INSULATION MATERIAL	....
2 CALCULATE RECIPROCAL OF U-VALUE THAT IS TO BE ACHIEVED (reciprocal = 1/U)	....
3 CARRY OUT U-VALUE CALCULATION OUTLINED PREVIOUSLY FOR WALL WITHOUT ADDED INSULATION (use worksheet no 1)	....
4 CALCULATE RECIPROCAL OF 3 ABOVE (reciprocal = 1/U)	....
5 SUBTRACT THE ANSWER TO 4 FROM THAT OF 2 (this is the thermal resistance that the wall lacks)	....
6 MULTIPLY THE RESULT OF 5 BY THE CONDUCTIVITY 'k' OF THE PROPOSED INSULATION MATERIAL, 1. (the answer is the required insulation thickness, in metres)	....

App 1.3 *Worksheet for calculating the maximum rate of heat loss (ie the outward heat flow rate) of an enclosed space such as a room or a building*

JOB TITLE				ENVELOPE OR ROOM
WORKSHEET NO				CALCULATION OF MAXIMUM RATE OF HEAT LOSS
FABRIC HEAT LOSS –				
1 FABRIC CONSTRUCTION	2 AREA m^2	3 U-VALUE W/m^2 degC	4 TEMPERATURE DIFFERENCE °C	5 RATE OF FABRIC HEAT LOSS (2 x 3 x 4) W
VENTILATION HEAT LOSS –				
VENTILATION LOCATION	NUMBER OF PEOPLE; ALTERNATIVELY VOLUME IN m^3	VENTILATION RATE PER m^3, OR PER PERSON	TEMPERATURE DIFFERENCE °C	RATE OF VENTILATION HEAT LOSS (2 x 3 x 4) W
TOTAL HEAT LOSS RATE (W)				

note – the watt is a measure of heat flow <u>rate</u>; for example one watt equals approximately 3600 joules per hour

App 1.4 *Worksheet for calculating the seasonal energy requirement of a building (based on BRE method)*

JOB TITLE			BUILDING	
WORKSHEET NO			CALCULATION OF HEAT GAINS (FOR SUBSEQUENT SEASONAL ENERGY REQUIREMENT CALCULATION)	
SOURCE	UNIT GAIN GJ			GAINS GJ
WINDOWS	ORIENTATION	AREA X m²	GAIN GJ	
	SOUTH		0.68	
	EAST & WEST		0.41	
	NORTH		0.25	
PEOPLE	1 GJ PER PERSON			
ELECTRICITY	BASED ON ESTIMATE OF CONSUMPTION. AVERAGE WATTAGE CONSUMED x 0.02 GIVES TOTAL HEAT GAIN IN GJ			
COOKING	GAS 6 GJ ELECTRICITY 4 GJ (UNLESS INCLUDED ABOVE)			
WATER HEATING	2GJ			
OTHER				
TOTAL GAINS (GJ)				

App 1.5 *Worksheet for calculating peak summer temperature in a room, stage one. This sheet takes the calculation as far as establishing the mean internal temperature*

JOB TITLE				
WORKSHEET NO			CALCULATION OF PEAK ENVIRONMENTAL TEMPERATURE - STAGE ONE.	

MEAN CONDITIONS : STAGE ONE

GAINS

SOLAR

SELECT DAILY MEAN SOLAR GAIN	X SOLAR GAIN FACTOR FOR GLAZING	X WINDOW AREA m^2	= MEAN SOLAR GAIN W

(1)

OTHER

SOURCES	RATE W	X DURATION hours	÷ 24	= DAILY MEAN GAIN W
LIGHTING (installed wattage)				
OCCUPANTS (No x Rate/Occupant)				
			TOTAL............................	

(2)

TOTAL

MEAN SOLAR GAIN (1)	+ MEAN DAILY GAIN (2)	= TOTAL MEAN GAIN W

(3)

LOSSES

FABRIC

EXTERNAL ELEMENT	AREA m^2	X U-VALUE W/m² degC	= RATE OF LOSS W/degC
WINDOW			
WALL			
		TOTAL............................	

(4)

VENTILATION RATE air changes/h table XVIII	X ROOM VOLUME m^3	X 0.33	= MEAN VENTILATION HEAT TRANSFER RATE W/degC

(5)

VENTILATION

FOR RATES OF VENTILATION OVER 2/hour COMPLETE :-				
1 ÷ answer to 5	= (a)	0.21 ÷ TOTAL AREA OF INTERNAL SURFACES m^2		= (b)
1 ÷		0.21 ÷		
(a) + (b)	= (c)	1 ÷ (c)	= MEAN VENTILATION HEAT TRANSFER RATE W/degC	
		1 ÷		

(6)

MEAN ENVIRONMENTAL TEMPERATURE

FABRIC LOSS (4)	+ VENTILATION LOSS (5) except if more than 2 changes/hour, then use (6)	= TOTAL RATE OF HEAT LOSS W/degC

(7)

TOTAL GAIN (3) W	÷ TOTAL LOSS (7) W/degC	= SUBTOTAL	+ MEAN EXT TEMP °C	= MEAN INTERNAL ENVIRONMENTAL TEMPERATURE °C

(8)

App 1.6 *Worksheet for establishing peak summer temperature, stage two and stage three. Stage two establishes the variations in internal temperature from the mean (which was calculated in the previous worksheet); and stage three establishes the peak temperature*

STAGE TWO: VARIATION FROM MEAN

SOLAR — VARIATION

PEAK INTENSITY SOLAR RADIAT'N W/m^2	DAILY MEAN - SOLAR INT. W/m^2	EFFECTIVE = PEAK INPUT	AREA OF X GLASS m^2	= SUBTOTAL	ALTERNATING X SOLAR GAIN FACTOR	EFFECTIVE = GAIN SWING W	
							(9)

CASUAL GAIN VARIATION — PEAK

CASUAL GAINS AT PEAK HOURS	RATE W	
TOTAL		(10)

CASUAL GAIN VARIATION — VARIATION

PEAK (10)	- MEAN (2)	= CASUAL GAIN VARIATION W	
			(11)

AIR TEMP — VARIATION

AREA OF GLAZING m^2	U-VALUE OF X GLAZING	HEAT = TRANSMITTED W/ degC	(5) OR USE + (6) IF MORE THAN 2/h	FABRIC PLUS = VENTILATION LOSSES	EXTERNAL AIR X TEMP SWING degC	AIR TEMP = INPUT VARIATION W	
							(12)

TOTAL — INPUT VARIATION

SOLAR VARIATION (9)	+ CASUAL VARIATION (11)	+ AIR VARIATION (12)	= TOTAL VARIATION W	
				(13)

INTERNAL ENVIRONMENTAL TEMPERATURE SWING — AREA X ADMITTANCE

INTERNAL SURFACE	AREA m^2	ADMITTANCE FACTOR FOR X CONSTRUCTION	AREA X ADMITTANCE = W/ degC	
		TOTAL		(14)

INTERNAL ENVIRONMENTAL TEMPERATURE SWING — SWING

AREA X ADMITTANCE (14)	+ VENTILATION GAINS (5) OR (6)	= SUBTOTAL	
			(15)

TOTAL SWING IN EFFECTIVE INPUT (13)	÷ (AREA X ADMITTANCE) + VENTILATION GAINS (15)	= SWING IN INTERNAL ENVIRONMENTAL TEMPERATURE degC	
			(16)

STAGE THREE — PEAK ENV TEMP

MEAN ENVIRONMENTAL TEMPERATURE (8)	+ SWING IN INTERNAL ENVIRONMENTAL TEMPERATURE (16)	= PEAK ENVIRONMENTAL TEMPERATURE degC

APPENDIX 2: CONVERSION CHARTS

Note on SI Units

SI Units are used throughout this book. This system was established in 1954 after a long period of progressive development of rational and coherent units. Many countries have adopted the system. It came into effect for the construction industry in the United Kingdom in 1971.

Traditionally systems of measurement have grown up employing many different units not rationally related. In the measurement of volume bushels, pecks, pints, gallons, cubic feet, cubic yards and a number of other units have been used simultaneously involving complex conversions for many otherwise simple calculations. In the field of heat energy the calorific values of different fuels were expressed in different units which rendered comparisons and calculations extremely difficult

eg	Coal	BTU per pound
	Oil	BTU per gallon
	Gas	Therms per cubic foot
	Electricity	Units (Board of Trade Units = 1kW hour)

The SI system provides a single unit, the joule(J), to be used for energy of all sorts and a single unit of volume, the cubic metre (m)). The table above becomes simplified to:

Coal	joules per kilogram
Oil	joules per m³
Gas	joules per m³
Electricity	joules per KWh

The simplification is apparent.

The SI system is based on six basic units:

Phenomenon	Unit	Symbol
Length	Metre	m
Mass	Kilogramme	kg
Time	Second	s
Electric Current	Ampere	A
Temperature	Degree Kelvin	°K
Luminous Intensity	Candela	cd

While the °K is used for absolute temperatures the degree celsius, °C, is still used for customary use. Degrees K and degrees C are identical in terms of temperature intervals.

From these basic and supplementary units the remainder of the units necessary for measurements are derived:

Area from length	$m \times m = m^2$
Volume from length	$m \times m \times m = m^3$
Velocity from length and time	m per s = m/s

Some derived units have special symbols:

Phenomenon	Unit	Symbol	Basic units involved
Frequency	Hertz	Hz	1 Hz = 1 cycle per second
Force	Newton	N	1N = 1 kg m/s
Energy (including quantity of heat)	joule	J	1J = 1 Nm
Power	Watt	W	1W = 1 J/s

Unwieldy decimals and the use of so-called 'scientific notation' are avoided by the use of named prefixes to specify multiples and sub-multiples of SI units. The multiples identify variations in number by a factor of 1000, eg 1000 joules is expressed as 1 kilo Joule (1000J = 1kJ)

SI multiples and sub-multiples

Factor		Prefix	
		name	symbol
one million million	10^{12}	tera	T
one thousand million	10^{9}	giga	G
one million	10^{6}	mega	M
one thousand	10^{3}	kilo	K
one thousandth	10^{-3}	milli	m
one millionth	10^{-6}	micro	μ
one thousand millionth	10^{-9}	nano	n
one million millionth	10^{-12}	pico	p

Phenomenon		Traditional unit	Conversion factor (trad unit × conv fact = SI Unit)	SI Unit	
				symbol	description
Space	length	foot	0·31	m	metre
		inch	0·025	m	metre
	area	square inch	645·2	mm²	square millimetre
		square inch	0·000645	m²	square metre
		square foot	0·09	m²	square metre
		square yard	0·84	m²	square metre
Volume		cubic inch	0·0000164	m³	cubic metre
		cubic foot	0·028	,,	,, ,,
		cubic yard	0·76	,,	,, ,,
		gallon UK	0·0045	,,	,, ,,
		gallon US	0·0038	,,	,, ,,
Mass	mass	ounce	0·0284	kg	kilogramme
		pound	0·45	,,	,,
		ton (short 2000 lb)	907·2	,,	,,
		ton (UK 2240 lb)	1016	,,	,,
Density		pounds per cubic foot	16·02	kg/m³	kilogramme per cubic metre
		pounds per gallon	9·98	,,	,,

Phenomenon		Traditional unit	Conversion factor (trad unit × conv fact = SI Unit)	SI Unit symbol	description
Temperature	customary temperature (level)	degree Farenheit	(°F—32) × 0·56	°C	degree Celsius
	temperature interval (range or difference)	degree Farenheit	0·56	°C	degree Celsius
Heat	quantity (energy)	British Thermal Unit BTU	1055	J	joule
		kilowatt hour	3·6	MJ	megajoule
		therm	105·5	GJ	giga joule
		calorie	4·187	J	joule
	flow rate (power)	BTU per hour	0·29	W	watt
		ton of refrigeration	3516	W	watt
	intensity of flow rate	BTU per square foot per hour	3·16	W/m^2	watts per square metre
Thermal properties	conductivity	BTU inch per hour square foot °F	0·14	W/m°C	watts per metre degree Celsius
	conductance	BTU per hour square foot °F	5·68	$W/m^2°C$	watts per metre squared degree Celsius
	resistivity	square foot °F per BTU inch $\left(\dfrac{1}{\text{conductivity}}\right)$	6·93	m°C/W	metre degree Celsius per Watt
	resistance	square foot hour °F per BTU inch	0·18	$m^2°C/W$	square metre degree Celsius per watt
	diffusivity	square foot per hour	0·000026	m/s	metre per second
	capacity per unit mass (specific heat capacity)	BTU/pound °F	4·19	kJ/kg°C	kilojoule per kilogramme degree Celsius
	capacity per unit volume (volumetric specific heat)	BTU per cubic foot °F	67·1	$kJ/m^3°C$	kilojoule per cubic metre degree Celsius
Calorific Value	weight basis	BTU per pound	2·32	kJ/kg	kilojoules per kilogramme
	volume basis	BTU per cubic foot	37·26	kJ/m^3	kilojoules per cubic metre
		BTU per gallon UK	5·99	kJ/m^3	,,
		BTU per gallon US	5·06	,,	,,
Latent Heat		BTU per pound	2·32	kJ/kg	kilojoules per kilogramme

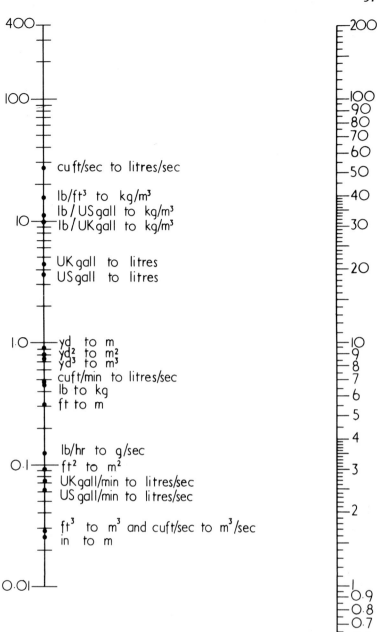

App 2.1 *Conversions: Imperial and US to SI—Dimensions, mass and flow.*

To convert imperial or US measures to SI unit values

1 *Select numerical value of Imperial or US quantity in the left-hand scale.*

2 *Select appropriate conversion point in the centre scale.*

3 *Project straight line from point found in the left-hand scale through conversion point in the centre scale to cut the right-hand scale.*

4 *Read off required SI unit value at point found in the right-hand scale.*

Note: a reverse process will effect conversion from SI to Imperial or US units.

Values beyond those shown in the scales may be converted. If too large, divide the value by a factor (eg 10, 100 or 1000) to bring it within the range shown in the scale, effect the conversion of the result and multiply the converted answer by the factor originally used.

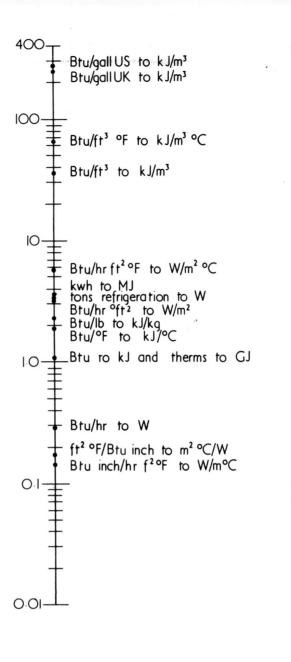

App 2.2 *Conversions: Imperial and US to SI—Heat, heat flow, thermal capacity, calorific value, etc. Use in the same way as* **App 2.1**.

temperature conversion scale
°F to °C temperature level

°F to °C temperature difference

App 2.3 *Temperatures require different conversions depending on whether the value being converted is a temperature difference (when a simple conversion ratio may be applied) or a specific temperature level when a correction to take account of the different zero points in the Fahrenheit and Centigrade scales must be applied in addition. Two scales are provided which enable °F to be converted to °C by selecting the appropriate value in °F on the left-hand graduation and reading off the corresponding value in °C on the right-hand graduation. If a specific temperature is being converted, the temperature level scale should be used. If a temperature difference is required, the temperature difference scale should be used.*

The reverse process may be used to convert °C to °F.

INDEX